Benjamin Wolf, PhD
George H. Snyder, PhD

Sustainable Soils
The Place of Organic Matter in Sustaining Soils and Their Productivity

Pre-publication
REVIEWS,
COMMENTARIES,
EVALUATIONS . . .

"*Sustainable Soils* by Benjamin Wolf and George Snyder offers a comprehensive account of the vital role of organic matter in the maintenance of sustainable agriculture. In this extremely valuable reference, the authors present a powerful case that sustainable agriculture cannot be attained unless sufficient organic matter is introduced into the system, and offer effective means of maximizing its usefulness. The chapters of this book describe in great detail factors that affect soil organic matter, the consequences of its loss, and the benefits of its increase, and emphasis throughout the book is appropriately placed on the importance of having sufficient soil organic matter rather than what can be expected if such levels are low.

The authors also present methods for increasing soil organic matter in a manner compatible with modern agriculture. They clearly demonstrate that a much better chance of achieving agriculture capable of sustaining maximum population is possible if a combination of maximum organic matter inputs and conservation tillage are used in conjunction with some inputs of chemical fertilizers and pest control.

Clearly written and illustrated, this book should be highly useful to all those who are interested in maintaining a long-term, profitable production that is environmentally friendly."

Nael M. El-Hout, PhD
*Senior Soil Scientist,
Research Department,
United States Sugar Corporation,
Clewiston, Florida*

More pre-publication
REVIEWS, COMMENTARIES, EVALUATIONS . . .

"*Sustainable Soils* is a well-written, comprehensive review of the role of organic matter in crop production. It will be useful for those in research and production who are faced with the present-day reality of crafting profitable cropping strategies using the very high-cost, man-made inputs on soils with impaired chemical and physical capabilities. The book deals with the scientific aspects of soil organic matter in a clear and understandable way, but growers and field technicians will be most pleased to find a wealth of information about equipment and strategies that are being used successfully today."

Theodore W. Winsberg, MS
Owner,
Green Cay Farms Incorporated

Food Products Press®
An Imprint of The Haworth Press, Inc.
New York • London • Oxford

NOTES FOR PROFESSIONAL LIBRARIANS AND LIBRARY USERS

This is an original book title published by Food Products Press®, an imprint of The Haworth Press, Inc. Unless otherwise noted in specific chapters with attribution, materials in this book have not been previously published elsewhere in any format or language.

CONSERVATION AND PRESERVATION NOTES

All books published by The Haworth Press, Inc. and its imprints are printed on certified pH neutral, acid free book grade paper. This paper meets the minimum requirements of American National Standard for Information Sciences-Permanence of Paper for Printed Material, ANSI Z39.48-1984.

Sustainable Soils
*The Place of Organic Matter
in Sustaining Soils
and Their Productivity*

FOOD PRODUCTS PRESS

Conservation Tillage in U.S. Agriculture: Environmental, Economic, and Policy Issues by Noel D. Uri

Mineral Nutrition of Crops: Fundamental Mechanisms and Implications edited by Zdenko Rengel

Wheat: Ecology and Physiology of Yield Determination edited by Emilio H. Satorre and Gustavo A. Slafer

Advances in Hemp Research edited by Paolo Ranalli

Dictionary of Plant Genetics and Molecular Biology by Gurbachan S. Miglani

Effects of Grain Marketing Systems on Grain Production: A Comparative Study of China and India by Zhang-Yue Zhou

A Produce Reference Guide to Fruits and Vegetables from Around the World: Nature's Harvest by Donald D. Heaton

Horticulture As Therapy: Principles and Practice by Sharon P. Simson and Martha C. Straus

Plant Alkaloids: A Guide to Their Discovery and Distribution by Robert F. Raffauf

Timber: Structure, Properties, Conversion, and Us, Seventh Edition by H. E. Desch, revised by J. M. Dinwoodie

Bramble Production: The Management and Marketing of Raspberries and Blackberries by Perry C. Crandall

Egg Science and Technology, Fourth Edition edited by William J. Stadelman and Owen J. Cotterill

Poultry Products Technology, Third Edition by George J. Mountney and Carmen R. Parkhurst

Seed Quality: Basic Mechanisms and Agricultural Implications edited by Amarjit S. Basra

Biodiversity and Pest Management in Agroecosystems by Miguel A. Altieri

Freshwater Crayfish Aquaculture in North America, Europe, and Australia: Families Astacidae, Cambarida, and Parastacidae by Jay V. Huner

Introduction to the General Principles of Aquaculture by Hans Ackefors, Jay V. Huner, and Mark Konikoff

Managing the Potato Production System by Bill B. Dean

Marketing Beef in Japan by William A. Kerr, Kurt S. Klein, Jill E. Hobbs, and Masaru Kagatsume

Understanding the Japanese Food and Agrimarket: A Multifaceted Opportunity edited by A. Desmond O'Rourke

Economics of Aquaculture by Curtis M. Jolly and Howard Clonts

Glossary of Vital Terms for the Home Gardener by Robert E. Gough

The Highbush Blueberry and Its Management by Robert E. Gough

The Fertile Triangle: The Interrelationship of Air, Water, and Nutrients in Maximizing Soil Productivity by Benjamin Wolf

Sustainable Soils
The Place of Organic Matter in Sustaining Soils and Their Productivity

Benjamin Wolf, PhD
George H. Snyder, PhD

Food Products Press®
An Imprint of The Haworth Press, Inc.
New York • London • Oxford

Published by

Food Products Press®, an imprint of The Haworth Press, Inc., 10 Alice Street, Binghamton, NY 13904-1580.

© 2003 by The Haworth Press, Inc. All rights reserved. No part of this work may be reproduced or utilized in any form or by any means, electronic or mechanical, including photocopying, microfilm, and recording, or by any information storage and retrieval system, without permission in writing from the publisher. Printed in the United States of America.

Cover design by Marylouise E. Doyle.

Library of Congress Cataloging-in-Publication Data

Wolf, Benjamin, 1913-
    Sustainable soils : the place of organic matter in sustaining soils and their productivity / Benjamin Wolf, George Snyder.
      p. cm.
Includes bibliographical references (p.) and index.
    ISBN 1-56022-916-0 (hard : alk. paper)—ISBN 1-56022-917-9 (soft : alk. paper)
1. Humus. 2. Conservation tillage. I. Snyder, George H. (George Heft), 1939- II. Title.
    S592.8 .W65 2003
    631.4'51—dc21

2002028819

CONTENTS

Foreword ix
Donald C. Reicosky

Preface xiii

Acknowledgments xix

Chapter 1. Intensive Agriculture and Food Production 1

Problems with Intensive Agriculture 2
Correcting Problems Associated with Intensive Agriculture 3
Alternative Agriculture 8
Sustainable Agriculture 11
Sustainable Soil Science 12
Increasing Organic Matter, Sustainable Agriculture, and the Need for Inputs 15

Chapter 2. Basic Concepts of Organic Matter 19

Decomposition of Organic Matter 19
SOM and Humus 23
Environmental Conditions Affecting Decomposition 26
Extent of Decomposition 33
Losses of SOM and Crop Yields 33
Increasing Soil Organic Matter 35
Negative Aspects of Organic Matter Additions 37

Chapter 3. Organic Matter As a Source of Nutrients 45

Organic Matter and Nutrient Supply 46
Nutrients Released from Organic Matter 48
Nutrient Release 54
Indirect Effects of Organic Matter on Nutrient Availability 57
Accounting for Nutrients in Organic Matter 64

Chapter 4. Physical Qualities — 79

Soil Structure — 80
Porosity — 84
Bulk Density — 91
Compaction — 93
Infiltration and Percolation — 95
Aggregate Stability — 98
Moisture-Holding Capacity — 99
Erosion — 102

Chapter 5. Biological Effects of Organic Matter — 105

Soil Organisms and Their Functions — 105
Processes Vital to Soil Health — 118
Controlling Pests by Harnessing Soil Organisms and Organic Matter — 125

Chapter 6. Adding Organic Matter — 137

Growing Organic Matter in Place — 137
Plant Residues — 162

Chapter 7. Adding Organic Matter Not Grown in Place — 169

Manure — 170
Compost — 178
Loamless Composts — 183
Sewage Effluents and Biosolids — 187
Peats — 191
High-Analysis Animal and Other Wastes — 192

Chapter 8. Placement of Organic Matter — 199

Incorporating OM — 199
Plastic Mulch — 200
Natural Organic Mulches — 201
Types of Natural Organic Mulch — 205
Deleterious Effects of Organic Mulches — 217
Providing Protection from Pests by Using Organic Mulches — 224

Chapter 9. Conservation Tillage — 227

- Increasing Soil Organic Matter — 227
- Other Benefits of Reduced Tillage — 228
- Reduction in Costs — 234
- No-Tillage — 234
- Reduced Tillage — 237
- Pest Problems with Conservation Tillage — 241
- Corrective Measures — 246

Chapter 10. Changes Brought About by Conservation Tillage — 249

- Cultural Methods — 249
- Fertility Changes — 256
- Pests — 263
- Plant Diseases — 267
- Nematodes — 272
- Weeds — 273
- Miscellaneous Pests — 283

Chapter 11. Putting It All Together: Combining Organic Matter Additions with Conservation Tillage — 285

- Basic Approaches — 285
- Adding Organic Matter — 288
- Supplying Needed Nutrients — 297
- Controlling Pests — 298
- Adding Water — 305
- Equipment for Conservation Tillage — 307

Summary — 313

Appendix 1. Common and Botanical Names of Plants — 319

Appendix 2. Abbreviations and Symbols Used in This Book — 327

Appendix 3. Useful Conversion Factors and Data — 331

Bibliography — 335

Index — 347

ABOUT THE AUTHORS

Benjamin Wolf, PhD, has over 60 years of research and consulting experience in soil, plant, and water analysis and crop production. Dr. Wolf developed practical methods of soil and plant analysis, greatly improving the determination of boron, showed that rapid soil tests can be used to evaluate fertilizer and lime requirements, used plant analyses to diagnose nutritional problems as early as 1945, and used the soil nitrogen test as a basis for applying nitrogen fertilizers as early as the mid-1940s. The improved rapid soil and plant analyses became the basis of one of the first independent consultation services in the United States, which is continued to this date, making it the oldest of its type in the country.

Dr. Wolf introduced micronutrients into fertilizer programs in several states and foreign countries, introduced foliar feeding for cooperative growers as early as 1949, introduced the practice of fertigation to cooperative growers in 1952, and was largely responsible for the development of the cut flower business in Colombia. Most of Dr. Wolf's present consulting work is done in Guatemala, Honduras, Costa Rica, and the Dominican Republic. He has written four previous books and over 40 articles in trade magazines and scientific journals.

George H. Snyder, PhD, is Distinguished Professor of Soil and Water Science at The University of Florida's Everglades Research and Education Center in Belle Glade. He has been a speaker/presenter at over 100 conferences and symposia, and has over 275 published articles. Dr. Snyder has been elected Fellow of the American Society of Agronomy and the Soil Science Society of America. He served as President of the Soil and Crop Science Society of Florida, and as Treasurer of the International Turfgrass Society. His research interests have ranged from wetland crop—and particularly rice—production on organic soils to production and environmental issues of turfgrass such as fertigation, nutrient requirements, supply, and leaching, as well as pesticide fate, transport, and risk assessment.

Foreword

Intensive agriculture has adequately met food production needs in the United States, but not without some negative impacts. Many soils in the United States have lost between 30 and 50 percent of the soil organic matter (SOM) present at the start of agricultural production. An extreme case is the organic soils of south Florida, where drainage and intensive tillage have caused loss of SOM that has resulted in three meters of soil subsidence. Ever since Jethro Tull (1674-1741), English agriculturist and inventor of the agricultural seeder, extolled the virtues of thoroughly tilling the soil to provide a mellow seedbed, there have been great efforts to develop and utilize machinery that would completely turn and/or mix crop residue and soil. The development of plows with iron tips, and later all-steel moldboard plows or plow disks, led to a thorough job of fracturing and turning the soil with considerable mixing. The development of the rototiller provided the ultimate in mixing soil and crop residue. These equipment developments, while initially beneficial for mining nutrients from SOM, have caused a serious decline in SOM that needs to be reevaluated and reversed.

The early benefits of intensive tillage, as it reduced surface compaction and greatly improved the available nutrient content of soils, were so dramatic that it greatly encouraged the development of machinery to do an ever better job. Plow contests championed more complete tillage, doing it better, deeper, and faster than ever before. In view of our present understanding of tillage practices, it is ironic that such contests are still in vogue, as indicated by a recent advertisement for such a contest painted on an Irish barn (see Figure 1). These events point to the social need for better education and communication about tillage impacts on SOM addressed in this book.

Our current knowledge equating tillage with adverse environmental responses has been developing for a few years, prompted by the early writings of Sir A. Howard (*An Agricultural Testament,* 1943, Oxford University Press, London), and E. H. Faulkner (*Plowman's Folly,* 1943, University of Oklahoma Press, Norman, Oklahoma). The con-

FIGURE 1. Advertisement for a plowing contest observed on an Irish barn. (Photo by G. H. Snyder.)

cept of reduced tillage for increasing or maintaining SOM and soil structure, and for decreasing energy requirements, has been investigated and slowly developed by research workers in state universities, the USDA, and at the Rodale Research Center (Kutztown, Pennsylvania). The combined work has greatly stimulated the use of conservation and reduced tillage, which has been accepted by segments of the farming community. These developments have led to the concept of conservation agriculture as a system of sustainable crop production.

Conservation agriculture has evolved to include more than just conservation tillage. Conservation agriculture represents a system that, when used appropriately, can provide a truly sustainable form of crop production. This concept does not describe a single operation or technique such as tillage, but an entire cropping system and a different way to understand interactions in a farming system. One of the main benefits of conservation agriculture is altering the amount and type of tillage in a production system, which reduces the loss of SOM, nutrients, and water as well as lowering input energy costs, time, and labor relative to more intensive conventional tillage. It has the potential for reducing erosion from wind and water, and for improving soil water status; soil biological, chemical, and physical properties and processes; and environmental quality. It further represents a cropping system that acts as a sink for carbon dioxide and allows a reduction of the emission of other pollutants to the atmosphere.

Notwithstanding considerable progress, much more has to be done to make SOM management in conservation agriculture the keystone of sustainable agriculture to produce sufficient food and fiber for present and near-future populations. Reduced tillage needs to be utilized on much larger acreage, but it also must be combined with increased additions of organic matter (OM) or crop residues properly placed as surface mulch. At times, some intensive inputs (quick-acting fertilizers and knock-down pesticides) must be used to compensate for the limitations of OM in providing optimum levels of nutrients and controlling pests.

Soil "life" plays a major role in many essential processes that determine nutrient and water availability and recycling and thus impact agricultural productivity. Crop residues and SOM are the primary food sources for living organisms in the soil and thus must be managed to be sustainable. With time, soil life takes over the functions of traditional soil tillage, which is loosening soil and mixing the soil and residue components. Biopores formed provide pathways for air and water flow important for crop production. In addition, the increased soil biological activity creates a stable soil structure through the accumulation of SOM. Intensive tillage destroys soil ecological integrity and limits the microbes' effectiveness in nutrient cycling. Thus, the most desirable form of tillage is not to till at all (zero tillage), which leaves a protective blanket of leaves, stems, and stalks from the previous crop on the soil surface for SOM enhancement.

While we learn more about SOM storage and its central role in direct environmental benefits, we must understand the secondary environmental benefits and what they mean to production agriculture. To minimize environmental impact, research results support that we must minimize the volume of soil disturbed and only till the soil volume necessary to get an effective seedbed and leave the remainder of the soil undisturbed to conserve water and SOM, and to minimize erosion and CO_2 loss. Increasing SOM storage can increase infiltration, increase fertility, decrease wind and water erosion, minimize compaction, enhance water quality, decrease carbon emissions, impede pesticide movement, and enhance environmental quality. Incorporating SOM storage in conservation planning demonstrates concern for our global resources. Communicating these concerns to the general public will lead to improved understanding of the social responsibility of agriculture. Agriculture using reduced tillage techniques can be of benefit to society and be viewed as both feeding and "greening" the world. The next step is to make society aware of the benefits and the social costs. This book, *Sustainable Soils: The Place of Organic Matter in Sustaining Soils and Their Productivity,* advocates the combination of increased OM additions, proper placement of OM as a mulch, and reduced tillage as outlined in a combined system (COM). This system includes (1) adding organic matter either by growing it in place, bringing it in, or using rotations and cover crops that increase organic matter; and (2) the use of conservation tillage to avoid burying the organic matter or destroying it by increasing aeration. Sufficient OM inputs to produce maximum economic yields will go far in furthering sustainable soils, agriculture, and environmental quality. Acceptance of the COM system described here is aided by the supporting data for increasing OM additions, proper placement (mulching), supplying needed nutrients, controlling pests, and providing suitable seeding equipment compatible with reduced tillage. Scientists must challenge social and political leaders with a new vision of conservation agriculture, one that balances our need to increase productivity of food, fiber, and biofuels, protects natural ecosystems, and improves the quality of our air, soils, and waters to improve our quality of life. The authors have done an admirable job in this regard, demonstrating the numerous benefits of reduced tillage and SOM accumulation.

Donald C. Reicosky
Soil Scientist, USDA-ARS
North Central Soil Conservation Research Lab
Morris, Minnesota

Preface

Abandoned farmland, that is what it had become. Once the sandy south Florida fields had been grasslands supporting herds of breeding cattle. But sugarcane provided a more steady income, so the fields were converted to a crop for which most of the aboveground plant material is either burned off or hauled away to the mill, leaving little to be returned to the soil as organic matter (OM). Over the years, in spite of good plant varieties, irrigation, fertilizers, and pesticides, production declined to the point where even sugarcane could no longer be profitably grown. The fields were abandoned.

Rice turned out to be the crop that made the fields productive again. The fields were planted to rice and flooded for months. Organic matter losses were checked by the flooding, and with rice, only the grain was removed. The straw was disked back into the soil, providing much-needed OM. Now sugarcane could be grown once again, and the rice-sugarcane rotation become a standard practice for improving and maintaining the fertility of these soils.

While crop production soared during the twentieth century with the advent of modern, intensive agricultural practices, soil organic matter (SOM) often declined appreciably. These losses have made farming more difficult, thereby requiring the use of even more intensive practices. Today, however, a new awareness of the importance of SOM both for crop production and to combat increases in atmospheric carbon dioxide is increasing. In this book we explore the factors that affect SOM, the consequences of its loss, and the benefits of its increase. Methods are described for increasing soil organic matter in a manner compatible with modern agriculture. Recent research in sustainable agriculture has indicated that it may be difficult to achieve unless a modified soil science becomes an integral part of the program. In the various alternative programs examined, it is quite apparent that the objectives of sustainable agriculture cannot be attained, at least in the foreseeable future, unless sufficient OM is introduced into the system and means are provided to maximize its usefulness.

To clarify matters, it might help to consider newly introduced organic materials, which are still recognizable, as organic matter, while that which has undergone sufficient decomposition and is no longer recognizable is known as soil organic matter. The latter, coming from many different sources that are quite different in composition and appearance, results in a rather similar complex known as humus. Although more stable than the OM from which it is derived, humus still is transitory in nature and will break down, the rapidity increasing with ample oxygen, good moisture levels, and elevated temperatures. The chemical formula for humus, which has similar appearance and composition regardless of its source, has not been determined, probably because of its variability as it decomposes.

The importance of adding sufficient OM and the need for sufficient SOM for long-term crop production has been accentuated many times over the years for both authors. For the senior author, the importance of both was highlighted rather early in his professional career. Soon after taking a new position at Seabrook Farms Company in southern New Jersey in 1941, he was asked to determine why a crop failure could be expected about half the time on about 1,000 acres of this 15,000 acre farm devoted primarily to vegetables. The failures occurred on land, mostly sandy loams, that had been devoted to intensive vegetable production for the longest period. Most of the farm had originally been in a mixed livestock (mostly dairy cattle) and small grain and potato production. The oldest division had been in intensive vegetable production for about 30 years, while the newer divisions were converted to intensive production in the previous five to ten years.

Until about the late 1920s, crops in this older division received large amounts of horse manure, which came from the livery stables in Philadelphia about 30 miles away. Yields and quality of vegetable crops were excellent, and remained so for a few years after manure applications were discontinued. Despite increased fertilizer use, yields began to decline in the early 1930s, with periodic crop failures appearing in the mid-1930s. Failures were more apt to occur if dry periods occurred as the cash crop (English peas) was maturing.

Extensive soil analyses showed that the only apparent difference between the areas prone to crop failure and the rest of the farm was that the soils from the poor area had SOM contents of 1.0 percent or less, while the other soils had 1.3 to 2.0 percent, with the average slightly over 1.5 percent.

It was assumed that much of the crop failure on soils with little SOM was due to poor water relationships usually existing in such soils, since nutrient levels were about the same in both areas, and the degree of failure dropped dramatically after irrigation was introduced.

The extent of crop failure prior to the introduction of irrigation was probably accentuated by a poor cover crop policy. Soon after livery stable manure was no longer available, a short study lasting only a couple of years was made to determine whether cover crops could compensate for the lack of manure. Unfortunately, only nonleguminous crops of either rye or annual ryegrass were used as test crops, and the dry matter production of both was very small because of the short time span available to grow them. The cover crops were planted in late October or early November after the fall crop of spinach or broccoli was harvested and was plowed under in March prior to planting English peas. Because the experiment showed no differences in yield of vegetable crops grown with or without these poor cover crops, all use of cover crops was abandoned. It was concluded that this abandonment harmed crops grown on soil with low SOM more than those grown on soil with more SOM due to differences in erosion losses usually exhibited between these different soils.

The study was a great learning experience. It emphasized the need to take OM additions and SOM levels into consideration for long-term successful crop production. Failure to do so could be costly. Also, the cost often is not realized for some time because SOM loss may be so slow that its effects may be overlooked for some time.

Somewhat similar results explained why serious problems of muskmelon yields in Guatemala were occurring after about six years of continuous high-intensity farming using plastic mulch, methyl bromide fumigation, and liquid fertilization through drip irrigation. The primary manifestation of problems was the poor movement of water or fertilizer solutions throughout the bed. Again, most problems were associated with soils containing about 1 percent of soil organic matter. At the start of intensive cultivation, yields were excellent and the organic matter content was about 2 percent. Problems have decreased dramatically since cover crops, primarily sorghum, have been grown between crops of melons.

As another example, in the southern United States, golf greens composed of medium and coarse sands are vegetatively planted with bermudagrass, which spreads over the surface due to its stolon and rhizome production (Figure 2). When no organic matter is included in the root zone mix, coverage can be quite slow. But with as little as 0.5 to 1.0 percent (by weight) included organic matter (sphagnum peat), coverage is greatly accelerated. A small amount of organic matter provides easily visible increases in the rate of bermudagrass coverage.

In the course of consulting work by the senior author extending over the past 50 years, the proportion of poor crops due to insufficient SOM as observed by him has increased. If problems caused by pests are excluded, those caused by insufficient or excess nutrients, poor soil pH, or excess salts were the most common during the early years. Probably due to better nutrient and pH control through soil and plant analyses, such problems appear to be less serious now. But problems apparently due to insufficient SOM appear to be increasing. The problems which we believe are primarily due to the lack of SOM are (1) compact soils; (2) poor water infiltration, storage, and movement; and (3) greater numbers of soilborne diseases.

FIGURE 2. Much faster establishment and spread of bermudagrass sprigs was obtained when they were planted in quartz sand with approximately 0.5 percent (by weight) added peat (right) as opposed to sprigging into sand without peat (left). (Photo courtesy of Raymond H. Snyder.)

Other examples of low SOM adversely affecting crop production are presented in appropriate chapters, but the book emphasizes the importance of having sufficient SOM rather than what we can expect if levels are low. Some of the problems of intensive agriculture and methods to correct them along with the place of SOM in the maintenance of a sustainable agriculture are presented in Chapter 1. A basic description of SOM is furnished in Chapter 2. The importance of having sufficient SOM is presented in a series of chapters—the value of OM and SOM for supplying nutrients in Chapter 3; their contribution to good soil physical properties and their influence on the infiltration, storage, and movement of water as well as the effects on erosion in Chapter 4; and the part played in maintaining soil biological organisms and processes in Chapter 5. The sources for adding OM are presented in Chapters 6 and 7. Maximizing the effects of organic matter is discussed in two chapters dealing with the use of mulch (Chapter 8) and reduced tillage (Chapter 9). Adjusting cultural practices induced by reduced or conservation tillage is discussed in Chapter 10. The final chapter (11) deals with maximizing the combination of adding large quantities of OM primarily as mulch with conservation tillage (COM) and what cultural changes are necessary as we bring these two methods of increasing SOM together. It also indicates that a much better chance of achieving an agriculture capable of sustaining a maximum population is possible if we use the combination of maximum OM inputs, mulch, and conservation tillage with some use of intensive inputs of chemical fertilizers and pest controls.

This book should be highly useful to advanced students in soils, crop production, agronomy, and horticulture. It should be invaluable to growers who are interested in maintaining long-term profitable production that is environmentally friendly. It can be helpful in guiding consultants, extension personnel, and tradespeople who sell agricultural products by helping growers to select such procedures that will benefit all of us in the long run.

Acknowledgments

The authors acknowledge the help of the following, who provided substantial assistance in the book's preparation:

- Kathleen L. Krawchuk, Coordinator of Academic Support Services, University of Florida, Everglades Research and Education Center, IFAS, Belle Glade, Florida, who supplied books, reprints, and a number of references that enhanced many of the covered subjects.
- Norman Harrison, Computer Specialist, also at IFAS, Belle Glade, who furnished valuable assistance with graphics.
- Rick Reed, Coffee County Extension Service, Douglas, Georgia, for information about compost preparation at Douglas.
- Dr. Sharrad C. Phatak, Department of Horticulture, University of Georgia, Athens, for information gained on a tour of crops grown in eastern Georgia with conservation tillage.
- Dr. Ronald D. Morse, Department of Horticulture, Virginia Polytechnic Institute and State University, Blacksburg, Virginia, for the photo of a transplanter developed at the University capable of planting through crop residues. The authors are also indebted to Dr. Morse for other photos and text dealing with growing crops in killed mulch.
- Other photos of conservation tillage equipment were supplied by Adrian Puig, Everglades Farm Equipment, Belle Glade, Florida, and Matthew A. Weinheimer, Marketing Manager, John Deere Des Moines Works.

The senior author is indebted to David N. Warren, President of Central American Produce, who has been willing to modify an intensive system for growing melons in Guatemala in order to improve soil organic matter that had dropped to a level low enough to affect water infiltration. Thus far, including cover crops of sorghum and corn between melon crops has provided extra organic matter, a habitat for beneficial insects, and improved melon production. The studies are continuing to make full use of a recommended organic matter program by finding cover crops better suited for the area and determining practical approaches for planting the melon crop through the residues of cover crops left as mulch.

Chapter 1

Intensive Agriculture and Food Production

Over the past 75 years, modern agriculture has brought about tremendous increases in yields of most crops, due to a combination of several changes in agricultural practices. The more important ones are widespread use of genetically improved crops, increased use of fertilizers, the general application of pesticides to limit damage from insects, fungi, and nematodes, and greater use of irrigation and drainage to improve water relationships.

The two world wars were responsible for development of a number of the chemicals now being used as fertilizers or pesticides. Synthetic nitrogen capability built up during the wars supplied very inexpensive nitrogen that could be used in the form of anhydrous ammonia, nitrogen solutions, ammonium nitrate, or urea, or it could be combined with phosphates to produce dry mono- or diammonium phosphate. The increased production of acids, especially during World War II, made it possible to produce relatively cheap high-analysis superphosphates. Facilities used to manufacture various organic compounds for war use were turned to the production of a number of pesticides, with DDT, 2,4-D, and organophosphate insecticides being some of the more conspicuous examples.

Several other changes in this period increased the efficiency of farmers. The shift from man and animal power to that of machines enabled the grower to efficiently farm much larger acreage at lower costs. Developments in large-scale machinery and the changing economics of using such machinery along with the freeing of acreage devoted to maintenance of draft animals aided and abetted the huge development of monoculture. The newer machinery, capable of quickly working large acreage, was more adaptable to the cultivation of large fields. As a result, the hedgerows between small fields were eliminated. At the same time, some of the large machinery, with its considerable cost, was less adaptable to cultivating a variety of different crops in the same year. Also, a more rapid write-off of the cost of the machinery was possible if only one crop was cultivated, justifying the use of monoculture.

Much of the increase in production has been due to expanded use of irrigation. Large areas of the West were made highly productive as huge dams supplied sufficient water for arid and semiarid areas. Irrigation became available in other areas, some of which have normal amounts of yearly rainfall but may lack sufficient quantities for certain periods. The expansion of irrigation, especially in areas that receive normal yearly rainfall, was accelerated by development of portable lightweight aluminum pipe that could be easily moved from field to field, and self-propelled units that automatically irrigated large acreage.

The huge increase in food production supplied the needs of a rapidly expanding population. Food production in the United States, a leading country in utilizing the modern concepts, remained sufficient as the population increased from about 115,000,000 in 1925 to over 275,000,000 in 2000, and even managed to have large surplus amounts for export. The efficiency of production is exemplified by the fact that over 10,000,000 farm workers were necessary to produce the food in 1925, and only about 1,000,000 were needed in 1998.

PROBLEMS WITH INTENSIVE AGRICULTURE

The huge increase in food production has not come without some costs. Some of the problems were apparent rather early. The use of large-scale machinery favoring increased field size, accompanied by elimination of hedgerows and the sods used to support draft animals, led to considerable erosion by wind and water. The great Dust Bowl calamity of the 1930s, while initiated by extreme drought, was made much worse by rampant tillage of much marginal land and the elimination of sods and trees that could have mitigated the disaster.

The adaptation of monoculture has tended to provide a less stable agriculture because the single crop makes it more vulnerable to market fluctuations. Also, monoculture appears to aggravate instability because it has led to more pest problems and greater soil degradation due at least partially to the fact that continuous cultivation to produce one crop reduces soil organic matter (SOM).

Other problems with crop production based largely on unlimited chemical inputs have arisen, calling the wisdom of continuing the process into question. One of the more challenging objections to the system is that the unbridled use of chemicals is detrimental to both animal and human health.

Rachel Carson, in her book *Silent Spring* (1962), caught the attention of the nation by claiming that the widespread use of agricultural chemicals was harmful to life. She singled out the effect of DDT on bird populations, with the prediction that continued use of pesticides might eliminate species of birds. The implication was that if these materials were so deadly as to eliminate entire species of birds, what might they be doing to humans?

The use of artificial fertilizers has received its share of criticism. The degree varies with the source. Proponents of organic farming, who use only animal manure and composted materials, certain religious sects that resist modern devices and practices, and those who have a mistaken belief that plants can only safely use organic fertilizers are the loudest in their criticisms. Some of these critics claim that artificial fertilizers are harmful to the soil and to both plant and animal life. It is argued that foods grown with these fertilizers, if not actually toxic to life, are less nutritious. But some criticism has come from ecologists and agricultural scientists who have been concerned with adverse soil physical changes and the harm to the environment caused by the unlimited use of artificial fertilizers.

Another serious criticism of agricultural production propelled by unlimited inputs is that irrigation, which is now used on about one-third of productive land, cannot be used indefinitely. The history of irrigation is replete with failures due to soil waterlogging, salinization of soils or the water source, and depletion of water supplies.

CORRECTING PROBLEMS ASSOCIATED WITH INTENSIVE AGRICULTURE

Various measures have been used over the years to overcome or mitigate some of the ill effects or deficits common to intensive agriculture. Some of the more important changes aimed at answering the most severe criticisms are (1) soil conservation as promulgated by the Soil Conservation Service in order to reduce erosion, (2) various legislative acts that regulate the use of pesticides so as to reduce or eliminate their harmful effects, (3) enrichment of fertilizers with micronutrients, (4) better fertilizer sources and methods of applying fertilizers to reduce contamination of water sources, and (5) introduction of drip irrigation to lessen water use combined with education on how irrigation can be used to reduce the problems of waterlogging, salinization, and overuse.

Erosion

Spurred in part by damage resulting from the Dust Bowl, there has been a consistent effort to reduce soil erosion by wind and water. Legislation passed in 1935 established the Soil Conservation Service in the Department of Agriculture, which worked closely with many state representatives to produce the Standard State Soil Conservation Districts Law. This was passed in 1937, and has become a modus operandi for establishing and administering erosion control projects. Although the Soil Conservation Service has not eliminated soil erosion, it and the related work of many individuals and organizations have made the likelihood of another Dust Bowl rather remote.

Pesticides

Legal regulation of pesticides began early in the twentieth century, and expanded greatly with the passage in 1947 of FIFRA (Federal Insecticide, Fungicide, and Rodenticide Act) and the changes made in 1954 to the Federal Food, Drug, and Cosmetic Act. FIFRA required that before any pesticide could move across state lines it had to be true to its label in terms of its effectiveness, safety, and contents of both active and inert ingredients. The changes to the Federal Food Drug and Cosmetic Act established tolerances of registered pesticides that could be allowed on foods. Further regulation of pesticides was promoted to a great degree by the concern raised by Rachael Carson and like-minded individuals, many of whom were motivated by her book.

As a result of such concerns, a major change in regulation took place in 1972 and again in 1975 as Congress completely revised FIFRA by an amendment to the act, which established the Environmental Protection Agency (EPA). Under the new act, the EPA became the major arbiter of pesticide regulation, although allowances were made for close cooperation between the federal government and the states. The act allowed the EPA to register pesticides based on extensive data on efficiency, toxic properties, and residue provided by the supplier; establish general- and restricted-use pesticides; set standards for commercial and private certified applicators to handle restricted pesticides; make it illegal to use a pesticide in a manner contrary to its label; and suspend or cancel registrations of pesticides if later evidence proves the pesticide to be ineffective or harmful to humans or environment.

The regulation of pesticides has eliminated much of the risk in using them. Establishment of restricted-use pesticides and regulation of their application, setting tolerances on foods, label information, and the removal of pesticides that are unduly harmful have done much to alleviate the problems of using pesticides. Over the years, some of the many pesticides considered unusually toxic (arsenic and mercury compounds, organochlorine and organophosphate insecticides) have been eliminated or their use greatly curtailed.

Although these changes were very helpful in reducing the inherent dangers, there remains considerable criticism of pesticide use. Some of the criticism is almost cultist, with little scientific evidence to verify the fears, but other issues, such as the possibility of hormone disruption in humans, remain potential problems, primarily because there are no definite answers yet.

Fertilizers

Much of the early criticism of commercial fertilizers was directed against their inability to supply essential nutrients for plants and animals. Much of this was justified, as commercial fertilizers lost several essential elements when they were developed to high-analysis grades. Very early fertilizers consisted primarily of various animal and plant wastes. Animal bones and even human bones taken from battlefields were the principle source of phosphorus (P); animal manures and guano (bird droppings) were major sources of nitrogen (N); wood ashes and wool wastes supplied much of the potassium (K). Later, fertilizers might have ammonium sulfate or sodium nitrate as sources of nitrogen; basic slag, superphosphate, or precipitate phosphorus as sources of phosphorus; and kainite, potassium chloride, manure salts, potassium sulfate, or sulfate of potash-magnesia as sources of potassium, but these chemicals were often used with relatively high organic sources of nitrogen, such as castor pomace, cottonseed meal, dried blood, fish scrap, and tankage (animal residues).

It is important to note that most of these organic materials carry a wide range of elements in addition to the primary element. Even the nonorganic materials carry small amounts or traces of other elements. Largely unknown at the time, these secondary and trace elements, such as boron (B), copper (C), iron (Fe), magnesium (Mg), molybdenum (Mo), manganese (Mn), sulfur (S), and zinc (Zn) were helping to maintain good yields of many crops.

Two developments precipitated a major crisis in crop fertilization and lent credence to the assumption that inorganic fertilizers were inferior to the natural organics. The natural fertilizer sources high in nitrogen were diverted to pet foods as this market outbid agriculture for the proteins (nitrogen sources) contained in animal and plant wastes, robbing agriculture of much needed trace elements, which are now known as micronutrients. Concurrently, the fertilizer industry, in its effort to bring even less expensive fertilizer sources to farmers, made higher-analysis fertilizers by eliminating much of the low-analysis materials. For example, items such as manure salts (24.5 percent K_2O) and kainite (14 to 22 percent K_2O) were abandoned as sources of potash, and the higher-analysis potassium chloride (60 percent K_2O) was used instead. Later, purification of potassium chloride provided 62 percent K_2O, which practically eliminated any possibility of supplying any element other than potassium and chlorine.

The net result of elimination of the natural organics as sources of nitrogen and the abandonment of low-analysis materials for fertilizers, coupled with reduced use of organics (manures, sods, cover crops) as agriculture went to monoculture and commercial fertilizers, was the appearance of an alarming number of crop problems. The problems were most severe on sandier soils in humid climates as the relatively small natural supply of micronutrients in these soils was lost to leaching and crop removal.

Fortunately, scientific studies revealed that many of these problems were due to insufficient micronutrients. Although the essentiality of the major elements (nitrogen, phosphorus, and potassium) and the secondary elements (calcium, magnesium, and sulfur) was discovered in the 1800s, the importance of most micronutrients was not discovered until the twentieth century—boron in 1926, copper in 1931, manganese in 1922, molybdenum in 1939, and zinc in 1926 (the essentiality of iron had been demonstrated in the 1860s) (Glass 1989).

As a result of these studies, various micronutrients were added to commercial fertilizers. The additions were guided by the use of soil and plant analyses, the practical use of which was largely developed in the period 1930 to 1960. The scientific addition of micronutrients to commercial fertilizers eliminated much of the nutritional problems which had appeared with the introduction of high-analysis grades. Today, there is no reason why crops grown with commercial fertilizers should be nutritionally inferior to those grown with organic fertilizers.

Other criticisms of inorganic fertilizers range from pollution, namely eutrophication of lakes and reservoirs, to the degradation of the soil itself. Fertilizers have been associated with pollution of both groundwater and surface water. Soluble fertilizers are rapidly leached into groundwater. The primary problem appears to come from nitrates, but other nutrients, including heavy metals, may add to the problems. Large quantities of nitrogen and phosphorus may also move into surface waters and cause serious problems with algae bloom that can result in fish kills. Nitrogen usually moves as a solution in runoff, but it and phosphorus can move with eroded soil.

Although there is no question that high use of fertilizers contributes to pollution of ground and surface waters, this type of pollution is also aggravated by nutrients from organic sources, such as manure, biosolids, compost, and decaying vegetation of many kinds. Overloading the soil with organic or inorganic fertilizers can cause problems, but some of these can be mitigated by (1) testing the soil for available nitrogen (nitrate and ammoniacal nitrogen), which helps limit applications to ranges handled by intended crops; (2) applying larger portions of nitrogen as coated fertilizers or in less soluble forms; and (3) keeping cover crops on the soil at all times to limit erosion and leaching.

Thus far agriculture has not done a good job of eliminating nutrient pollution of surface and groundwater. Seldom is nitrogen applied based on tests for available nitrogen. Seldom are both nitrate and ammoniacal forms determined and even less frequently are enough determinations made during the growing season to fully evaluate the need. Slow-release forms, such as coated fertilizers, have been used to a limited extent, primarily because of cost. Most of these fertilizers are used for high-priced items, such as flowers, vegetables, or herbs. The use of cover crops year round has been increasing, but much more will have to be done before we can markedly reduce pollution of ground and surface waters. The authors believe that greatly increased attention to building SOM through reduced tillage and added OM, keeping most of it as mulch, probably can greatly reduce the problems of nutrient pollution.

The accusation that inorganic fertilizers "poison" the soil appears to be largely misdirected. Although excessive use of artificial fertilizers may in rare instances "poison" the soil because of toxic levels of element(s) or buildup of salts, such events are rather rare today with modern use of soil and leaf analysis. Rather, the "poisoning" or loss of productivity of some soils with intensive agriculture and heavy use of artificial fertilizer appears to be due to the effects of extensive tillage with monoculture upon SOM. Actually, additions of fertilizer are usually associated with an increase in plant residues (OM), but the increase evidently is not enough to offset the losses of SOM that usually occur because of the accelerated decomposition and erosion resulting from the increased tillage and lack of crop rotation that are integral parts of intensive agriculture. The loss of SOM is manifested by many changes in soil properties—most of which have a negative impact on soil productivity.

ALTERNATIVE AGRICULTURE

Despite the many changes introduced to solve problems related to intensive agriculture, there have been a series of movements to utilize alternative forms of agriculture, which may be more sustainable. Spurred by high energy costs, severe soil erosion, and severe drains on nonrenewable sources, the American Society of Agronomy (ASA), the Soil Science Society of America (SSSA), and the Crop Science Society of America (CSSA) met in Atlanta, Georgia, November 29 through December 3, 1981, to reevaluate whether organic farming might be a basis for developing a sustainable agriculture.

For various reasons, some of which have been cited already, a significant number of farmers have either refused to go the intensive route or, having tried intensive farming, have found it less profitable or desirable and reverted to organic farming. Over the years, as problems arose with intensive agriculture, there have been attempts to find the answers in organic farming.

The definition of organic farming may vary with different growers. It evidently is not, as many people think, simply conventional agriculture without synthetic fertilizers and pesticides. It appears to more closely resemble a definition originally supplied by the USDA Study Team on Organic Farming, cited by Eggert and Kahrmann (1984):

> Organic farming is a production system which avoids or largely excludes the use of synthetically compounded fertilizers, pesticides, growth regulators, and livestock feed additives. To the maximum extent feasible, organic farming systems rely upon crop rotation, crop residues, animal manure, legumes, green manures, off-farm organic wastes, mechanical cultivation, mineral bearing rocks, and aspects of biological pest control to maintain soil productivity and tilth, to supply nutrients and to control insects, weeds and other pests. (pp. 98-99)

After considerable review of the papers presented at the Atlanta meeting, Elliott et al. (1984) stated that organic farming could well contribute to a more sustainable agriculture, "and that our crop production systems can best be served by combining principles of organic agriculture with those of conventional and conservation tillage agriculture" (p. 188).

Another look at alternative agriculture, which has gained appreciable attention from scientific agriculture, was made by the committee appointed by the Board on Agriculture in 1984 under the auspices of the National Research Council. The committee was formed primarily because of financial problems apparent in the 1980s, when over 200,000 farms went bankrupt. Alternate forms of agriculture were to be considered as means of reducing excess costs. At the same time, it was felt that the environmental consequences of intensive farming needed to be examined. Essentially, the focus was directed to "three goals: (1) keeping U.S. farm exports competitive; (2) cutting production costs; and (3) reducing the environmental consequences of farming" (Robbins 1989, p. v).

The committee examined alternative management methods as indicated in 11 case studies of alternative methods of farming, which differed from conventional farming methods to different degrees. The committee had difficulty in comparing costs and benefits of intensive agriculture with that of alternative programs, indicating that true evaluation of the two systems would have to await further scientific study. Nevertheless, the committee came to these conclusions:

1. Federal government price support policies tolerate or encourage "unrealistically high yield goals, inefficient fertilizer and pesticide use, and unsustainable use of land and water" and by so doing discourage alternative practices such as rotations, application of conservation principles, or reduction of pesticides.
2. Alternative farming systems can be productive and profitable often without price supports.
3. Alternative programs are not well-defined but rather consist of various management practices that tend to reduce costs while protecting "health and environmental quality and enhance beneficial biological interactions and natural processes."
4. Most alternative farming systems tend to "use less synthetic chemical pesticides, fertilizers and antibiotics per unit of production than comparable conventional farms." Such reduction does not necessarily reduce per-acre crop yields or returns.
5. Successful alternative systems usually require more time, information, and better-trained labor as well as better management skills than most conventional farms. (Robbins 1989, pp. 8-10)

Some of the practices listed for alternative agriculture have been adopted by a number of intensive farms, and there is a good possibility that more of these practices will be adopted, at least for certain conditions or regions, as more growers become aware of their long-term benefits. Some of the more promising practices are

1. methods to reduce application of pesticides by use of IPM (integrated pest management), selection of pest-resistant cultivars, use of insect parasites and predators to combat pests, and insect pheromones to disrupt their mating;
2. crop rotations or use of polyculture (growing more than one crop) for better pest control;
3. reducing nitrogen needs by including a legume in the rotation;

4. further reduction in fertilizer use by proper timing and placement of fertilizers and by incorporating, where possible, leguminous forages and cover crops; and
5. reduced tillage to conserve SOM, reduce erosion, and limit costs.

SUSTAINABLE AGRICULTURE

The search for alternative to intensive input use has been accelerated because it is considered nonsustainable. Various definitions for sustainable agriculture have been proposed. From a grower's point of view, any system is sustainable if it keeps the farm in business. From a longer-term view, a more politically sensitive definition may be one quoted by Letey (1994) taken from Section 1404 of the Natural Agricultural Research, Extension and Policy Act of 1977, as amended by Section 1603 of the FACT Act, which states that sustainable agriculture is

> an integrated system of plant and animal production practices having a site-specific application that will, over the long term: (i) satisfy human food and fiber needs; (ii) enhance environmental quality and the natural resource base upon which the agricultural economy depends; (iii) make the most efficient use of non-renewable resources and on-farm resources and incorporate, where appropriate, the natural biological cycles and controls; (iv) sustain the economic viability of farm operations; and (v) enhance the quality of life for farmers and society as a whole. (p. 23)

We have to agree with Letey that this definition is too complex and cumbersome. In addition, we find that some of it is rather vague. For example, what is "the most efficient use" or what constitutes an enhanced "quality of life for farmers and society as a whole"? A simpler definition offered by the Brundland Report (Brundland 1987) states that sustainability "meets the needs and aspirations of the present without compromising the ability of future generations to meet their own needs." But it, too, tends to be vague as to "aspirations." Perhaps it might be useful if we consider agriculture to be sustainable if the farmer's practices keep the farm in business while producing enough food and fiber for present generations without unduly affecting the quality of production facilities, substantially depleting nonrenewable sources, or polluting waters, so that subsequent growers will be able to produce sufficient food and fiber for future generations.

SUSTAINABLE SOIL SCIENCE

From our own experiences and the literature of many scientific studies, current agriculture is not sustainable unless the soil itself is sustained by various practices, or farming methods include a soil science that enhances soil fertility and limits its loss by wind or water erosion. While it is possible to produce various crops without soil by hydroponics and aeroponics, economic considerations limit the production to high-priced crops. For a long time to come, the bulk of the food and fiber necessary for human maintenance will have to be grown on soils.

The Place of Organic Matter in Providing Sustainable Soil

Sustainable agriculture is not possible without sustainable soil. But soil cannot be sustained without satisfactory SOM, which in turn is largely dependent on OM additions and how they are handled. The dependence of sustainable agriculture on organic matter originates from the many beneficial effects of both OM and SOM forms.

Soil organic matter consists of a wide variety of plant and animal tissues in various stages of decomposition, from that which is slightly decayed to that which is no longer recognizable. The decaying materials, coming from many sources that can be quite different in composition and appearance, result in a rather similar complex known as humus. Although more stable than the organic materials from which it is derived, humus is transitory in nature and will break down, albeit very slowly, the rapidity increasing in soils with ample oxygen and good moisture levels and at elevated temperatures. The chemical composition of humus has not been determined, probably because of its variability as it decomposes.

Both OM and SOM are largely responsible for soil formation and development. The energy from both sources as they decay supports a vast number of microorganisms (bacteria, fungi, actinomycetes). These organisms are constantly modifying the rocks from which soil is formed, releasing nutrients for plants. The energy released from organic matter also benefits larger organisms (mites, earthworms, and insects) that intimately mix the fine rock fragments with organic matter, greatly hastening the decomposition of the rocks and speeding soil formation.

By maintaining large numbers of diverse organisms, organic matter helps maintain a healthy balance between beneficial and disease organisms. But many more benefits derive from organic matter, because organic matter supplies the energy for a number of useful processes carried out in the soil, without which it would become difficult if not impossible to provide satisfactory production.

Some of the more important processes vital for agriculture and supported by the energy derived from organic matter are

1. the breakdown of organic matter, which releases a number of elements present in organic forms, and largely unavailable to plants, to inorganic forms which are readily absorbed by plants;
2. the nurturing of both symbiotic and free-living organisms that convert atmospheric nitrogen (N_2), which is not available to plants, into readily available forms of ammoniacal (NH_4^+) and nitrate (NO_3^-) nitrogen;
3. the sustenance of mycorrhiza fungi that help keep phosphorus in an available form; and
4. the support of bacteria, actinomycetes, and some filamentous fungi that aid in the formation of cements so essential for binding individual small soil particles into aggregates or peds.

The aggregates formed greatly improve soil structure and lessen soil bulk density. Improved soil structure markedly improves crop production by

1. improving water infiltration, thereby lessening soil erosion and increasing the amount of available water;
2. increasing air porosity, which allows better movement of air and water in soils;
3. providing an ideal environment for beneficial microorganisms and plant roots;
4. aiding good tilth, which is necessary for ease in soil preparation as well as rapid development of seedlings and plants; and

5. reducing erosion by aiding infiltration, and favoring certain soil characteristics that aid in the movement of suitable amounts of air and water, without which crop production is not possible.

Organic matter aids soil productivity in several other ways. It:

1. increases cation exchange capacity (CEC), allowing better retention of ammonium nitrogen, potassium, calcium, and magnesium);
2. provides for chelation of several micronutrients, which helps keep them available;
3. helps keep phosphorus available, particularly at both high and low pH values;
4. buffers soil, limiting rapid changes in pH or salt content that can occur with addition of various chemicals;
5. decreases dispersion of soil by raindrops or irrigation and thus lessens surface crusts and compaction; and
6. lessens changes in soil temperatures, which could interfere with nutrient availability and plant survival.

The various benefits of organic matter are considered in greater detail in the chapters that follow. The decomposition of organic matter, which supplies energy for various organisms and the processes that they carry out is considered in greater detail in Chapter 2. Additional evaluation of OM and SOM is discussed in several chapters. Chapter 3 covers the nutrients supplied by various forms of organic matter and the role they can play in crop production. Chapter 4 deals with the physical changes in soils resulting from organic matter. Chapter 5 covers the many biological effects of organic matter and how these effects alter the problems caused by various pests.

The Importance of Organic Matter in Sustaining Agriculture

The importance of OM and SOM in crop production has been known for a long time, but its essentiality for sustainable agriculture has been realized more recently. The delay in assessing the essentiality of SOM has occurred for at least two reasons: (1) the decrease in SOM from a satisfactory level to one that is associated with problems of crop production often takes several years, especially in cooler climates; and (2) the introduction of new intensive inputs often compensates for some of the defects due to existing intensive methods. Replacing micronutrients in chemical fertilizers, discussed previously, is an example of how a new intensive input overcame the negative results obtained with use of commercial fertilizers of the time. Other common examples are (1) the use of soil fumigants, such as methyl bromide, to overcome increased soil pest problems, aggravated by lower SOM and monoculture; and (2) expanded use of irrigation, which alleviated a number of problems associated with poor water storage due to lack of SOM.

INCREASING ORGANIC MATTER, SUSTAINABLE AGRICULTURE, AND THE NEED FOR INPUTS

Much of the movement to a more sustainable agriculture implies that we need to abandon such intensive inputs as chemical fertilizers and pesticides. The increased application of organic matter could well decrease the need for these inputs, but from our observations, OM alone cannot compensate for the complete abandonment of these inputs for a number of crops, especially in warmer or tropical climates. Although such agriculture could well be sustainable in the sense that many soils would continue to produce fairly good yields with a minimum of pollution, it would be insufficient to sustain much of the world's burgeoning population.

Elimination of chemical fertilizers and pesticides from cropping systems that include animal production probably would not limit yields as much as it would on farms without animals. Manure produced on the farm can supply considerable amounts of organic matter. Also, tillage of forage and grain crops, usually grown on these farms, is less than for more intensive crops. Reducing tillage aids in the preservation of SOM by reducing the amount of oxygen in the soil. The lower level of oxygen slows the breakdown of SOM by microorganisms, because oxygen is necessary for respiration, by which they obtain energy.

Reduction of inputs is possible even without livestock if an increase in organic matter is accompanied by eliminating as much tillage as possible. The reduction in inputs can be greater on many soils if much of the added OM is retained on the surface as a mulch.

Combining additions of cover crops and/or residues with reduced tillage lessens erosion, provides for better water infiltration and storage, improves soil aeration, aids weed control, slows the destruction of SOM, and provides a more equitable environment for soil microorganisms, thereby aiding several vital soil processes and reducing insect and disease damage. The surface placement tends to reduce water erosion, soil compaction, and soil temperature, all of which can be beneficial for many soils with good drainage. Several methods of incorporating additional OM with reduced tillage and mulch use are discussed in Chapter 8.

But can we or should we abandon all intensive additions? Again from our observations, such abandonment could be very costly. Nutrients supplied by cover or other crops often are insufficient to supply the needs of high-nutrient-requiring crops. The deficits can be supplied by manure or compost, which can be satisfactory if produced on the farm, but become rather cost-prohibitive for many crops when they have to be hauled some distance. Supplying the deficit of 50 lb each of N, P, and K requires 500 lb of a 10-10-10 fertilizer, whereas it would take about 10,000 lb of manure or about 5,000-7,000 lb of compost to do the job. Handling and hauling these large amounts to supply nutrients adds an unnecessary expense, especially if organic matter additions from other sources (cover crops, crop residues) were already supplying satisfactory physical and microbiological needs of the soil.

Even if the amounts of total nutrients are sufficient, the amounts available during cool periods or after a leaching rain could readily warrant the addition of fast-acting artificial fertilizers. Commercial fast-acting fertilizers are also called for when wide carbon:nitrogen (C:N) ratios of OM delay the release of enough nitrogen to prevent the new crop from starting rapidly. Abandoning the use of commercial fertilizers in these situations can mean a substantial reduction in some crop yields.

Likewise, the abandonment of all synthetic pesticides at all times could be disastrous. Although we can greatly reduce the amounts used by adding enough organic matter, and combining these additions with the use of IPM and resistant cultivars, while providing suitable environments for predators, some quick-acting pesticides are needed for the occasional uncontrolled outbreak. The need for these pesticides is greater in warmer climates where such outbreaks can occur with greater frequency.

Therefore, as we see it, the use of sufficient OM can greatly reduce the need for some of the intensive inputs, while substantially reducing erosion and pollution. The reduction of inputs can be much greater on many soils if the extra OM is coupled with reduced tillage and the organic materials are applied as a mulch. But even with these procedures, there is a need for some quick-acting chemical fertilizers and pesticides.

Chapter 2

Basic Concepts of Organic Matter

Organic matter in soils can exist in the form of recently added materials (OM) or that which has decomposed beyond recognition (SOM). Some basic consideration of the forms of organic matter should be useful in understanding how both of these can be used to full advantage in maintaining a sustainable agriculture.

DECOMPOSITION OF ORGANIC MATTER

All organic materials undergo decomposition in soils. Recently added materials consisting of plant and animal remains (crop residues, cover crops, small animals, insects, microorganisms) usually decompose very rapidly when incorporated into soil. Materials left on the surface will decompose at a slower rate. Materials that have undergone some decomposition and are no longer recognizable also break down more slowly. The decomposition of newly added materials depends primarily upon the type of material, its age, particle size, and N content, but soil moisture, temperature, aeration, pH, and nutrient content also affect the decomposition rate of both OM and SOM.

Young, succulent plant materials break down most rapidly, whereas more mature materials decay more slowly. Leaves and vines decay more rapidly than stems or roots. Plant sugars, starches, amino acids, and some proteins, which are present in large amounts in young tissue, break down very quickly, followed by most hemicelluloses and, last, by the lignins. The slow decomposition of lignins and some hemicelluloses may explain, at least partly, the slow decomposition of older material, since these compounds are primarily present in older tissue.

Young succulent materials break down more quickly, not only because of higher levels of sugars, amino acids, and proteins, but also because of the narrow ratio of carbon to nitrogen. Young succulent materials have C:N ratios closer to those of microorganisms (10:1), allowing rapid decomposition without the addition of N. Usually, these materials will be rich in N and low in lignin and polyphenol concentrations (Handayanto et al. 1997). On the other hand, older materials are rich in lignins and polyphenols with less N. Materials with low N and high concentrations of lignins and polyphenols (wide C:N ratio) decompose more slowly.

The C:N ratio markedly influences the decomposition rate and the mineralization of N because N determines the growth and turnover of the microorganisms that mineralize organic C, but the C:N ratio alone is not always a perfect indicator of the rate of decomposition. Dead plant materials have N contents of 0.1 to 5 percent, the C:N ratios of which are in the range of 20:1 to 500:1, but animal and microbial tissues with their high protein contents have ratios less than 20:1. The C:N ratios of many different materials that may be added as OM to soils are given in Table 2.1.

Normally, we can expect very rapid decomposition of OM with C:N ratios of less than 20:1, sometimes with free ammonia being released since nitrogenous compounds are used as C sources. Unlike the decomposition of animal and microbial tissues with C:N ratios of less than 25:1, the decomposition rate of tissues having C:N ratios between 25:1 and 75:1 is variable. Plant materials that include green leaves from green manure crops or crop residues decompose very rapidly, while some legumes will break down slowly. The materials containing green leaves have high available C:N and P because tissues have not senesced, a condition that encourages decomposition by microorganisms. On the other hand, some legumes with relatively narrow C:N ratios (15:1 to 35:1) may decompose slowly because of complexing of protein by polyphenols. (Polyphenols alter the chemical nature of proteins in the complexing process by making them a much larger and different chemical entity that is more difficult for microorganisms to decompose.) In such cases, the ratio of N to polyphenols may be a better indicator of decomposition rate than the C:N ratio. Cereal and legume straws and annual crop litter that contain about 10 to 15 percent lignin have C:N ratios of 50 to 100 and their decomposition rates are more normal since the higher ratios reflect a

TABLE 2.1. Carbon/nitrogen (C:N) ratios of some organic materials

Material	C:N ratio
Slaughterhouse wastes	2:1
Soil bacteria	5:1
Soil actinomycetes	6:1
Soil fungi	10:1
Night soil	6:1-10:1
Soil humus	10:1
Sewage sludge	10:1-12:1
Alfalfa	12:1
Sweet clover, young	12:1
Grass clippings, young	12:1
Chicken manure, droppings	12:1
Alfalfa hay	15:1
Municipal garbage	15:1
Hog manure	17:1
Sheep manure	17:1
Chicken manure, litter	18:1
Sweet clover, mature	20:1
Grass clippings, fresh mature	20:1
Rotted manure	20:1
Bluegrass	30:1
Cattle manure	30:1
Bean straw	37:1
Horse manure	49:1
Leaves, fresh	40:1-80:1
Pine needles	45:1
Peat moss	58:1
Corn stalks	40:1-80:1
Highmoor peat	80:1
Straw, small grain	80:1
Timothy	80:1
Cotton gin trash	80:1

TABLE 2.1 *(continued)*

Material	C:N ratio
Corncobs	104:1
Red alder sawdust	135:1
Seaweed (kelp)	225:1
Hardwood sawdust	250:1
Douglas fir, sawdust	295:1
Douglas fir, old bark	295:1
Wheat straw	375:1
Pine sawdust	729:1

Source: Follett, R. H., L. S. Murphy, and R. L. Donahue. 1981. *Fertilizers and Soil Amendments.* Englewood Cliffs, NJ: Prentice-Hall, Inc.; Lorenz, O. A. and D. N. Maynard. 1988. *Knott's Handbook for Vegetable Crops,* Third Edition. New York: John Wiley and Sons.

lower N content rather than changes in C. The decomposition of woody materials with C:N ratios above 75:1 or 100:1 also can be variable. The rates are normally slow but vary depending upon the lignin content because lignin is resistant to microbial attack and can protect cell wall sugars, normally easily decomposed, from microbial decomposition. As a result, the ratio of lignin to N may be a better indicator of decomposition rate of woody materials than a simple C:N ratio (Heal et al. 1997).

Fragile versus Nonfragile

The rapidity with which plant residues undergo decomposition varies not only with the type of tissue, C:N ratio of the material, and such chemicals as polyphenols and lignins but also with the age of the material, size and thickness of the leaves and stems, whether these are damaged by harvesting machinery, whether leaves have fallen from stems prior to harvest, and the density of the residues. Based on some of these criteria, residues are classified as fragile (decompose rapidly) or nonfragile (resistant to decomposition) in Table 2.2.

TABLE 2.2. Classification of residues as to their persistence (nonfragile or fragile)

Nonfragile residues	Fragile residues
Barley*	Beans, dry
Buckwheat	Canola/rapeseed
Corn	Corn silage
Flaxseed	Cover crop, fall seeded
Forage seed	Flower seed
Forage silage	Grapes
Hay	Guar
alfalfa	Lentils
grass	Mint
legume	
Millet	Mustard
Oats*	Peanuts
Popcorn	Peas, dry
Rice	Potatoes
Rye*	Safflower
Sorghum	Sorghum, silage
Spelt*	Soybeans
Sugarcane	Sugar beets
Tobacco	Sunflowers
Triticale*	Vegetables
Wheat*	

Source: Farnsworth, R. L., E. Giles, R. W. Frazes, and D. Peterson. 1983. *The Residue Dimension.* Land and Water # 9. Cooperative Extension Service, University of Illinois at Urbana-Champaign.
* Consider as fragile if a rotary combine is used for harvesting.

SOM AND HUMUS

Decomposition of OM proceeds to a point when its composition is appreciably different from either the plant or animal residues from which it originated. The stable end products typically consist primarily of ligninlike complexes and proteins. The original plant materials have varying amounts of lignin, high carbohydrates, and low protein levels, while animal residues originally have high levels of proteins and no lignin. Both materials change appreciably as the stable complex of SOM is formed. The change in composition results as the readily decomposable carbohydrates and amino acids in the original materials are quickly disposed of, leaving the more resistant lignin with some slower-decomposing carbohydrates, tannins, waxes, resins, and chitins. Some proteins are bound by polyphenol fractions, especially those with very condensed tannins. The lignins and their derivatives are combined with certain uronic acids. The composition of SOM largely reflects these end products, but is modified by compounds synthesized by microorganisms and the remains of the microorganisms.

The relatively stable material resulting from these changes, also classified as humus, amounts to a major portion of SOM. Humus, or the semistable product of OM decomposition "is not a chemical individual but a mixture of substances varying under different conditions of formation, depending on such factors as the nature of the vegetation, nature and intensity of its decomposition, climatic conditions, and physico-mechanical and chemical soil properties" (Waksman 1938, p. 93). It also has been considered to be the remainder of the material left in the forms of complex humic substances and nonhumic compounds, after most of the carbon (60-80 percent) from OM has been lost as carbon dioxide (CO_2). Humus and the biomass of soil organisms make up the total residue (SOM). The nonhumic compounds consist primarily of polysaccharides, polyuronides, and acids (Brady and Weil 1999).

Humus is quite homogenous and has a composition of about 50 percent organic C and 5 percent N on a dry weight basis. This provides a C:N ratio of 10:1, which is much narrower than the original plant materials from which the bulk of SOM was derived.

In addition to C and N, humus contains about 0.5 percent P, 0.5 percent S, and small amounts of K, Ca, Mg, and micronutrients. The micronutrient content can be important for intensive agriculture systems that use high-analysis fertilizers that have not been fortified with micronutrients.

Various attempts to categorize humus as a chemical entity have failed. The molecular weight of humus can vary tremendously, depending to a good extent on mode of extraction. Despite considerable similarities, humus is of variable composition because it is derived from materials that may differ markedly in original composition, plus it contains various compounds released from the original OM that may be different as they are modified by different organisms involved in the decomposition process. Added to these variables are (1) synthesized compounds that may be markedly different because they are formed by different organisms or (2) by the same organisms that function under different conditions such as temperature, moisture, and oxygen content and (3) the remains of organisms that are quite different in composition. No single chemical structure has been accepted for humus, probably because of differences in the organisms attacking the OM, conditions under which decomposition has taken place, the state of decomposition, and the mode of sample preparation.

Despite these variations, several chemical properties are fairly common to humus. One of these is the presence of functional groups that probably account for some of the properties generally observed. The equal presence of carboxyl and hydroxyl groups appears to account for its near neutrality. Another is the relatively high exchange capacity, which is primarily dependent on the carboxyl groups, but this varies from about 170 meq/100 g of dry weight for the high molecular weight fractions and 500 meq/100 g for the lower molecular weight fractions.

Humus is resistant to microbial action and to acid hydrolysis but is easily degraded by oxidation. Tillage, by greatly increasing the amount of O_2 in the soil, is a major factor in its decomposition.

Our concept of humus has varied with time, but there is a general idea that humus consists of humic and nonhumic substances. Although the composition of these components varies depending upon the method of extraction, there is some consensus that the humic substances are the major part of the humus. They have been classified into three groups mainly based on solubility. Fulvic acid, with the lowest molecular weight, is light in color and most easily attacked by microorganisms; humic acid is medium in molecular weight and color and semiresistant to microbial action. Humin has the highest molecular weight and is most resistant to decomposition, probably because it is protected from microbial action as it is complexed by clay.

Humic acid, which has gotten considerable attention in recent years, acts as an acid only when a major portion of its cation exchange sites are saturated with hydrogen ions, but with little effect on soil pH since the acid is insoluble in water. The rest of the time it acts somewhat as a clay. The humates formed as humic acids are saturated with various cations and have different solubilities. Those saturated with monovalent cations (Na^+, K^+) are soluble, whereas those with multivalent elements (Ca^{2+}, Mg^{2+}, AL^{3+}, Mn^{2+}, Mn^{3+}) are insoluble, but the cations have little effect on the humic substances other than the effects on solubility.

Humic acids have been found to have several beneficial effects on plants. They evidently favor the availability of Fe and certain other micronutrients. The effect on Fe probably accounts for the increased growth of some groups of microorganisms observed with humic acids. The availability of Fe even in the presence of considerable P_2O_5 appears to be due to chelation of Fe by humic acids. The availability of P_2O_5 itself increases with the presence of humic acids, possibly as the acids prevent formation of aluminum phosphates. Humic acids also have been reported to have auxinlike effects or growth-promoting actions, tending to stimulate germination of seeds. Some of this growth effect may be due to increased permeability of plant membranes that is stimulated by humic acids. Overall, humic acids appear to increase the uptake of nutrients, particularly Fe. Although much of the metal uptake has been attributed to chelation effects, that of N and other nonmetal elements as well as the metals may be due to polyphenols from humic acids that act as respiratory catalysts. These stimulate enzyme systems and cell division that could account for better root systems and better nutrient uptake (Senn and Kingman 1973).

ENVIRONMENTAL CONDITIONS AFFECTING DECOMPOSITION

Soil Oxygen

The amount of soil oxygen (O_2) is one of the more important factors affecting the rate of decomposition of both OM and SOM. Microorganisms that are largely responsible for decomposition depend on O_2 for their respiration. Conditions that restrict soil O_2, such as excess water or soil compaction, slow decomposition and result in the accumulation of SOM. An example is the accumulation of SOM in bogs with the formation of peats or mucks, or in mineral soils with good drainage that have not been cultivated. Losses of SOM in virgin soils soon after they are open to cultivation also are primarily due to increased aeration as soils are tilled, although some of it may be due to reduced amounts of added OM as they are farmed. The better SOM levels with rotations (Table 2.3) are due in part to the reduced cultivation needed for closely planted small grains as compared to wider-spaced rows of corn.

Farming practices affect amounts of SOM primarily by varying the amounts of O_2 introduced in a soil through cultivation, although the extent of mixing OM with soil particles and the amount of OM added can also alter amounts of SOM. Implements that aerate the soil tend to hasten decomposition. The moldboard plow, which does a very complete job of mixing OM with soil and introducing large amounts of air deep in the soil, tends to deplete SOM more rapidly than some other soil preparation implements. As we will see in Chapter 9, reducing tillage tends to preserve SOM primarily because it limits the amount of air introduced into the soil.

Oxygen is also largely responsible for the differences in SOM commonly found in different soil textural classes. Sands, which allow the greatest amount of air because of high porosity associated with the large particles in their composition, tend to have the lowest SOM of all textural classes. Loams, usually having less porosity than sands, will have intermediate values, while clays, with the least porosity, tend to have the highest values. The values for loams and clays are modified by the degree of aggregation, which influences porosity and the amount of soil oxygen. If porosity is high in these soils due to good aggregation, the more favorable growing conditions due to in-

TABLE 2.3. The effect of rotations on the organic matter of some Ohio soils

	SOM (%)
Initial content	4.4
30 years continuous corn	1.6
30 years continuous wheat	2.8
Rotation of corn, oats, wheat, clover, timothy	3.4
Rotation of corn, wheat, clover	3.7

Source: Adapted from Salter, R. M. and T. C. Green. 1933. Factors affecting the accumulation and loss of nitrogen and organic carbon in cropped soils. *Journal of the American Society of Agronomy* 25:622. Reproduced by permission of the American Society of Agronomy, Madison, WI.

creased soil air could result in enough additional plant residues to more than overcome the extra loss attributed to increased microbial action.

Some of the differences in SOM noted between different textural class soils are due to differences in the amounts of organic materials produced on these different soils. The amounts of OM produced often reflect differences of moisture and nutrient contents in these textural classes. Sands usually produce the smallest amount because of limited moisture and nutrients.

The greater amount of SOM in soils of finer textures is not only the result of larger amounts of OM produced on these soils but is also affected by the smaller amount of O_2 available for decomposing the SOM. Less O_2 is available because larger amounts of it are consumed by microorganisms as they decompose the relatively larger amounts of organic matter produced, and smaller amounts of O_2 reenter these fine-textured soils.

This utilization of O_2 and release of CO_2 in the decomposition process can result in reduction of O_2 and an increase in CO_2 in the soil to levels that inhibit decomposition, particularly if OM is present in large amounts and/or there is poor gas exchange. The latter usually results when soils are too wet, too compact, or have large amounts of fine particles (silt and clay) that are dispersed. The lost O_2 is difficult to replace under these circumstances. The inhibition resulting from decreased O_2 and accumulation of CO_2 tends to hinder the decomposition of any remaining OM and may be disastrous to the primary crop because of O_2 shortages.

The changes in O_2 concentrations as decomposition proceeds are illustrated in Figure 2.1. Considerable O_2 is consumed, and large amounts of CO_2 are formed in the decomposition process. Figure 2.2 depicts CO_2 being measured as soils are tilled.

Atmospheric Carbon Dioxide

In the process of decomposition, about 60 to 80 percent of the C in the original material is lost as carbon dioxide. Greater C loss as CO_2 results in lower SOM, which can have negative effects on soil properties and, therefore, affect soil sustainability, but the evolved CO_2 released from the decomposition of organic matter into the air also adds to the burden of atmospheric CO_2, which may be excessive already

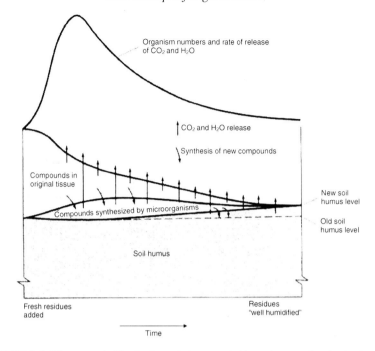

FIGURE 2.1. Diagrammic illustration of changes taking place as fresh plant residues are added to soil. (*Source:* Brady, N. C. 1990. *The Nature and Properties of Soils,* Tenth Edition. New York: Macmillan Publishing Company; London: Collier Macmillan Publishers. Reprinted by permission of Pearson Education, Inc., Upper Saddle River, NJ.)

due to the buring of fossil fuels. High levels of atmospheric CO_2 have become a concern to many scientists and public officials dealing with weather and policy matters, primarily because high levels may be responsible for the "greenhouse effect," which appears to cause global warming. It would appear that the burden of excess atmospheric CO_2 can be reduced appreciably by sequestering larger amounts of C in SOM.

As we will see in later chapters, larger amounts of SOM can be maintained if larger amounts of OM are added, and there is considerable reduction in tillage that reduces the amounts of O_2 in the soil. Maintaining larger amounts of SOM is a win-win policy because it provides for sustainability of soil and reduces somewhat the adverse effects of global warming.

FIGURE 2.2. A device for measuring CO_2 within a plastic chamber is being lowered over freshly tilled soil to document the effect of tillage on microbial activity and resultant CO_2 evolution. (Photo courtesy of Dr. Don Reicosky, USDA.)

Clay

The larger amounts of SOM commonly found in clay soils can also be attributed to the protective nature of clay as well as the reduced level of O_2 usually present in finer-textured soils. Clay appears to protect SOM from microbial action through a combination of (1) adsorption of SOM on clay surfaces, (2) entrapment of SOM in pores or aggregates derived from clays, and (3) the encirclement of SOM by clay particles. The preservation effect increases with the percentage of clay in different soil textural classes, being least in sands and increasing in the order loams-silts-clays.

This protective effect probably accounts for changes in soil C, an important constituent of SOM. Cultivation tends to lower C much more in sands than in silts. Instead of lowering C in clay soils as it does in sands and silts, cultivation may increase it (Dalal and Mayer 1986). The increase probably occurs because cultivation introduces enough O_2 to stimulate additional growth, which provides extra SOM on these normally oxygen-poor soils, while the protective action of the clay reduces decomposition of the SOM normally accelerated by cultivation.

Temperature

Temperature is another important factor affecting the decomposition rate of OM and SOM, which approximately doubles for about every 10°C (18°F) increase in the growing range of 13-35°C (45-95°F). The SOM contents in uncultivated soils of the northern plains are often more than 4.0 percent, whereas they are more commonly 2 percent or less in the southern plains. Microbial activity usually comes to an abrupt halt as soon as soils freeze. Although total organic matter produced on some northern soils is usually lower than in the warmer regions, the SOM levels of the northern soils tend to be greater because of restricted microbial activity for an appreciable part of the year.

Lower temperature also accounts for the lower amounts of organic materials in the form of crop residues or manure that need to be added to the soil to maintain existing SOM levels. In temperate zones, this amounts to about two tons of OM per acre per year, but only half of this is needed in cooler climates. In the tropics, where it is more difficult to maintain adequate SOM levels, higher SOM levels are obtained if the soil is cooled by shading, which occurs, for example, when certain varieties of coffee are grown.

Moisture

Moisture also affects the rate of decomposition. Whereas increases in soil temperature tend to decrease SOM, increases in soil moisture tend to increase SOM. Moisture levels satisfactory for crop growth are also ideal for organic matter decomposition, but SOM levels usually increase in this range, partly because of increased residues obtained with ideal moisture contents. If moisture levels are excessive for crop production, SOM levels will not drop and may even rise because O_2 is restricted as moisture displaces it in the soil pores. The restriction in decomposition due to insufficient O_2 helps to preserve SOM. The SOM can also increase when OM lost to crop plants is made up by plants that tolerate or even flourish with excessive moisture. On the other hand, SOM values usually are lower in dry climates without irrigation because of lower OM production.

Nutrient Levels and pH Values

Despite the fact that microbial activity capable of decomposing OM usually increases in soils with near-neutral pH and adequate crop nutrient values, SOM usually increases. Evidently, the increased amounts of residues, above and below ground, that result from better crop yields on soils with suitable pH values and nutrient content more than make up for the increased decomposition.

Nitrogen may well be a large factor in the increase of SOM usually noted in soils having higher nutrient values. Nitrogen applications for cotton were associated with increases in SOM over a 16-year period. Nitrogen application rates of 125-150 lb/acre increased SOM about 33 percent (0.3 percent SOM) compared to 0 nitrogen applications (Table 2.4).

Nitrogen applications also were associated with SOM increases in both conventional and reduced tillage used for continuous corn. Increasing nitrogen fertilization from 0 to 300 lb/acre resulted in sub-

TABLE 2.4. Influence of nitrogen rates on soil organic matter for continuous cotton

N rate (lb/acre)	Percent organic matter*		
	1973	1984	1989
0	0.66	0.69	0.9
25	0.69	0.75	1.0
50	0.66	0.89	1.1
75	0.69	0.89	1.1
100	0.67	0.91	1.1
125	0.68	1.02	1.2
150	0.68	0.93	1.2

Source: Maples, R. I. 1989-1990. Nitrogen can increase cotton yield and soil organic matter. *Better Crops with Plant Food* Winter. Atlanta, GA: Potash and Phosphate Institute. Reproduced by permission of the Potash and Phosphate Institute, Atlanta, GA.
*Organic matter content of upper six inches of a Loring soil

stantial increases in SOM in the 0-6 inch layer. With conventional tillage, the percentage of SOM increased with N applications up to 150 lb/acre, but under no tillage, percentage of SOM continued to increase with the 300 lb N rate (Sprague and Triplett 1986).

EXTENT OF DECOMPOSITION

Losses of SOM do not continue indefinitely over time but reach an equilibrium. At that point, losses are just equal to the new SOM derived from additional introduced OM. Losses in virgin soils are quite rapid at first, but then decline and finally stabilize as a new equilibrium is reached. Virgin soils in the corn belt lost about 25 percent in the first 20 years after being opened to cultivation, but about 10 percent in the next twenty years and only 7 percent in the next twenty years (Jenny 1941).

The rate of loss may be greater in warmer climates. Based on soil test data accumulated by the senior author, an area in Guatemala, which has a short rainy season of about four months, of grazing land with grass and small brush that had about 2 percent SOM at the start of cultivation was reduced to about 1 percent in a seven-year period of intensive cultivation. In that period, two to three crops of melons were grown each year, leaving little time for adding OM. The SOM did not drop below 1 percent and may have increased slightly in the next six years, but intensity of cropping was reduced slightly by limiting melon crops to two per year and growing sorghum as a cover crop for a period of about two months in the summer.

LOSSES OF SOM AND CROP YIELDS

The decomposition of SOM does not necessarily mean reduced crop yields. In fact, some decomposition of organic matter is essential for crop production. As we have seen in Chapter 1, the decomposition of organic matter provides energy for several vital processes, without which sustainability of agriculture would be difficult if not impossible to maintain. The decomposition of organic matter is not harmful unless losses due to decomposition are so much greater than the amounts added that certain important soil characteristics are negatively affected.

There will be no loss of productivity if additions of OM and the soil ecological system allow sufficient SOM to be retained to satisfactorily grow the intended crop. Losses usually do occur if SOM falls low enough to compromise certain soil values such as the degradation of soil structure, increased soil erosion or compaction, reduced infiltration, reduced movement or storage of water, poor gas exchange, or creation of an inhospitable environment for beneficial microorganisms.

To what point can SOM fall before crop yields are compromised? There have been various attempts to establish the minimal amount of SOM that can still provide good productivity. Sikoura and Stout (1996), in trying to determine a target level of C and N that would characterize "good quality" soil, concluded that such a target would have to depend on climatic region and soil texture. They gave an example that a fertile high-quality soil in the southwestern United States may have only 1 percent organic C, but a soil with the same amount of C in the Midwest might be considered of poor quality.

Soil texture evidently is a very important factor in determining minimal amounts of SOM that are compatible with good production, since the data cited earlier for Seabrook Farms in southern New Jersey indicated that appreciably lower amounts of SOM were associated with good yields of English peas on sandy loams than if the same crop was grown on loam soils (Wolf et al. 1985).

Climate, particularly rainfall, also is a factor in quantifying SOM levels needed to produce large crops. A big factor in establishing the level of 1 percent SOM as a bare minimum for sandy soils was the fact that failure to produce a sufficient crop to pay harvesting and processing costs was expected on sandy soils with less than 1 percent SOM about 50 percent of the years. The number of crop failures on these low-SOM soils was greatly reduced after irrigation was introduced. Although crop yields were lower on low-SOM soils than on sandy soils with higher SOM values, crop failures were almost eliminated and the differences between yields of crops grown on soils with less than 1 percent SOM and those with higher values were diminished.

Evidently, no definite value for minimal satisfactory SOM values can be given for all areas, as this appears to vary with the crop, soil textural class, climate (particularly rainfall), irrigation, and tillage methods. Very good productivity can be maintained if sufficient OM is added each year to maintain adequate porosity, allow sufficient water storage and movement, and limit erosion. Sands with large amounts of coarse particles usually have sufficient porosity allowing adequate air and water movement, but they do lack sufficient water storage. Introduction of irrigation (especially drip irrigation) for sands largely overcomes the more serious defects associated with low SOM values. The heavier soils (loams, silts, and clays) usually have satisfactory water storage, but may lack adequate air and water movement unless OM and SOM provide sufficient mucilages to aggregate the fine clay particles into peds. The peds provide better porosity, allowing more air and water movement. To maintain adequate production on heavier soils, enough OM must be added to at least maintain sufficient peds for satisfactory air and water exchange.

INCREASING SOIL ORGANIC MATTER

Organic Matter Additions

Ideally, SOM needs to be increased in many soils under intensive cultivation, but this is no easy task, and any long-lasting increases may be impossible to obtain by just increasing the amounts of OM applied. The amounts needed are huge. The 1 percent loss in the Guatamalan soils cited previously is equivalent to 20,000 lb of dry matter per acre. Since about 65 percent of the C in added OM is often lost as CO_2 during the decomposition process, leaving only 35 percent as SOM, about 57,000 lb of OM would have to be added per acre to replace the lost 1 percent. Even if it were economically feasible to add such large amounts, the addition could cause serious problems for the succeeding crop. As pointed out below, it is very difficult to apply large amounts of various organic materials without upsetting soil moisture and oxygen balance as well as posing serious physical problems. Also, there is a good chance that much of the added OM will be gone in a few years, unless substantial amounts are added each year or there is a marked change in cultivation practices.

There are benefits to adding OM even though SOM may not increase or may increase very little over the long term. The added OM helps to curtail erosion, improves water infiltration, provides nutrients, and provides a more favorable environment for microorganisms. Usually, saprophytic organisms (bacteria and fungi that live off decaying plant and animal tissues) substantially increase, providing competition with organisms capable of causing several soilborne diseases.

The increase in SOM with OM additions depends on the nature of OM added. Green manure or cover crops appear to add little SOM, although substantial benefits can be obtained from their use. Annual additions of manure tend to boost SOM if the SOM level has reached an equilibrium. If SOM is still in decline, manure additions reduce the rate of loss. Even with large annual applications of farmyard manure, the rate of SOM increase is very slow or nonexistent. Experiments conducted at Rothamsted, England, the oldest continually operating agricultural experiment station in the world, indicated that annual dressings of 14 tons of farmyard manure per acre decreased soil N loss under continuous cropping with small cereals about 27 percent less than accrued under continuous cropping without manure (Table 2.5). The 0.042 percent N saved by addition of manure is equivalent to about 0.84 percent SOM.

Change in Cultural Practices

Even with increases of added OM, change in SOM is rather slow unless there are marked changes in climate or cultural practices that limit oxygen introduction into soil. The rapid drop in SOM as virgin soils or longtime sods are opened for cultivation occurs primarily be-

TABLE 2.5. Loss of soil organic matter as indicated by nitrogen* loss over a 50-year period as affected by applications of complete fertilizer or farmyard manure

Crop	Complete fertilizer		Farmyard manure	
	N	Loss of N	N	Loss of N
	(%)		(%)	
Continuous barley	0.109	30	0.151	3
Continuous wheat	0.104	33	0.145	7

Source: Modification of data presented by Waksman, S. A. 1938. *Humus*. Baltimore, MD: The Williams and Wilkins Company. Reproduced by permission of Lippincott Williams & Wilkins ©1938.
*Multiply by 20 to obtain SOM.

cause soil O_2 markedly increases, allowing a marked increase in microbial action. The reduction in SOM slows with time and eventually ceases as the SOM reaches a new equilibrium, which reflects the new levels of OM introduced in the new system and the new levels of soil O_2. Assuming climate stays the same, the new equilibrium tends to remain constant until there is again a shift in composition of the soil environment or in the amounts of OM added.

Marked changes in cultural practices tend to change SOM more rapidly than the mere addition of manure, cover crops, sludge, or compost. Changes in cultural practices are effective because they tend to change the soil environment as well as often adding OM. The inclusion of sods, hay crops, or small grains in rotations is beneficial because the reduced cultivation with these crops tends to slow decomposition of OM and SOM, while the amount of OM added usually exceeds that obtained with wide-spaced row crops that require cultivation for weed control.

No-tillage or reduced tillage also allows increases in SOM, primarily because of lower soil O_2 content. The benefits of no-tillage and reduced tillage are covered in greater detail in Chapter 9.

NEGATIVE ASPECTS OF ORGANIC MATTER ADDITIONS

Although the addition of OM to soils is usually beneficial, there are occasions when additions can be harmful to succeeding crop(s). Most of the negative effects appear to be related to the use of such large amounts that they affect the soil environment (mostly oxygen and moisture), but occasionally it is the nature of the material that causes the problem. Some of the more important problems and possible solutions are briefly discussed here.

Excess Organic Matter

Oxygen and Gas Exchange

As has been pointed out, large amounts of CO_2 and shortages of O_2 resulting from the decomposition of OM can cause problems with crop productivity. Incorporating large amounts of OM can temporarily lessen O_2 contents, causing serious shortages that affect a plant's ability to carry on normal respiration, thereby limiting its ability to take up nutrients and water. The reduced level of O_2 results because greatly increased numbers of microorganisms, stimulated by the organic matter additions, take it up for their needs. In addition to reduced levels of O_2, the increased activity of microorganisms releases large amounts of CO_2, altering the composition of air within the pores. Altering pore air composition can slow the normal exchange of CO_2 for needed O_2, adversely affecting plant growth.

Oxygen shortages can be worsened by large additions of wet organic materials. The shortage of O_2 and excess of CO_2 induced by microorganisms as these materials are added is exacerbated by the high moisture levels of some added materials, particularly as they pack, limiting gas exchange and the amounts of available O_2. At times, the problem can be aggravated by methane and other undesirable gases produced under anaerobic conditions.

Dry Organic Materials

Incorporation of large additions of dry organic materials also may induce problems because these materials may cause temporary moisture shortages by absorbing large quantities of water and physically interfering with capillary movement. They may also prevent good contact of soil and seed, seriously affecting germination. The problem is usually worse if the OM is in relatively large pieces.

The negative effect of large OM additions will be overcome in time. If planting can be delayed, excellent crops from large OM additions can be expected once the materials have undergone considerable decomposition. Delaying planting until the OM is no longer recognizable and soil moisture and added OM have reached equilibrium usually provides a good growing medium. The process is hastened by plowing and disking to obtain a more uniform mixture of soil and OM. Excess disking that induces packing needs to be avoided.

Since plowing and disking increase soil O_2 and hasten destruction of the OM, a better answer to the problem is to avoid incorporation, leaving the OM on the surface as mulch. If enough time is available, the OM will break down sufficiently to avoid causing any problems planting through it, providing a proper coulter is used to cut through the trash. Amounts too great to cut through can be handled by pushing some of it aside prior to seeding.

Micronutrient Deficiencies

Large additions of OM may cause deficiencies of copper, boron, iron, and zinc. Copper deficiencies are quite common on organic soils unless considerable Cu is added, because Cu can be immobilized by organic matter, making it unavailable to plants. Although less common than Cu deficiency, B deficiency has been noted on organic soils, evidently due to the high adsorption of B by organic matter (Hue et al. 1988). Increasing levels of compost made from cattle manure added to a loess soil decreased B in the soil solution as indicated by lower amounts of boron in bell pepper leaves (Yermiyahu et al. 2001). The shortages of Fe and Zn caused by additions of OM do not appear to be caused by sorption on organic matter. These shortages, plus some Cu and B shortages, appear to be due to one or more of the following: (1) failure of the concentration of the elements in the soil solution to keep up with the greatly increased demand for these elements by the increased microbiological population as it is stimulated by the OM additions, (2) reduction of plant nutrient uptake due to a lower soil O_2 content induced by the tremendous increase in the numbers and activity of microorganisms, and (3) lower availability of these elements due to an increased pH resulting from large amounts of CO_2 produced by the increased microbiological population as it decomposes the OM. Carbon dioxide tends to increase bicarbonate content of high-pH soils, inducing iron deficiency in many plants.

Adverse Effects Due to Organic Matter Composition

Carbon:Nitrogen Ratios

As was seen previously and is covered more fully in Chapter 3, the release of nutrients from OM usually depends upon the C:N ratio of the added material. The release of nutrients by decomposition of OM with wide C:N ratios is usually accompanied by a reduction of soil N to meet the demand of the microorganisms, diminishing the amount available for plant growth. The problem of N deficit usually is aggravated by large additions of OM with wide C:N ratios.

Allelopathic Effects

Plants contain a number of substances, such as volatile terpenoids, quinones, thiocyanates, coumarins, tannins, flavonoids, and cyanogenic glucosides, which regulate density and limit competition from other plants by limiting germination of many seeds and restricting their growth. The effect is usually expressed as plant residues decompose, which may seriously impact the succeeding crop. The toxic effect is commonly expressed as corn follows corn, rye, or wheat, or when wheat follows rye or wheat. Subterranean clover and ryegrass, used as cover crops, can suppress succeeding crops of lettuce, broccoli, and tomatoes. Oats can suppress lettuce, cress, peas, wheat, rice, and timothy.

At times, the allelopathic effect can be beneficial to the succeeding crop. The value of several cover crops is enhanced because they have some herbicidal properties against a number of weeds. Some cover crops with allelopathic properties capable of suppressing weeds are black oats, hairy vetch, oats, rye, sorghum-sudangrass hybrids, subterranean clover, sweet clover, and woolypod vetch (Clark 1998). The allelopathic properties of rice against a number of weeds have been known for some time. Recent studies have shown that rice can affect the viability of subsequent broadleaf weeds, with some rice cultivars being more effective than others. Extracts of hulls are more potent than those of stem or leaf. The hull extracts are particularly effective against barnyard grass, which threatens to be a serious weed in direct-seeded rice culture (Ahn and Chang 2000).

Harmful allelopathic effects often can be materially reduced by delaying planting of the succeeding crop for a few weeks. This may allow sufficient time to decompose enough of the toxic substances to negate much of the harmful effect. Where the toxic substances persist for more than a few weeks, or the growing season is too short to allow for extensive delays, it is best to substitute an alternate cover crop if it is causing the problem or use an alternate primary crop that is not affected. If reduced tillage is being used, the affected crop may be planted, providing planter attachments can be used to remove most of the offending residue from the seed row. By so doing, harmful substances are moved away from the germinating seed or early developing roots, which usually are highly sensitive to these substances.

Composting OM prior to its use tends to overcome the allelopathic effect. Composting to avoid damage from these substances is a viable approach for certain high-priced crops, pot cultures, and urban use. The use of wood wastes for pot cultures often creates some restrictions on the crop unless the wood wastes are composted. Composting wood wastes overcomes the adverse effects of a wide C:N ratio as well as any allelopathic effects. The wood wastes may be composted separately or in various mixtures with sand, peat, or perlite.

Heavy Metals

Sufficient heavy metals may be introduced with some organic materials, such as sewage sludge (biosolids) or composts made from urban wastes, to affect succeeding crops. The elements cadmium (Cd), chromium (Cr), cobalt (Co), lead (Pb), and nickel (Ni) may be present in sufficient quantities in these organic materials to cause problems for the succeeding crop or make the crop unsuitable for human consumption.

Poultry wastes can contain arsenic (As), cobalt, copper, iron, manganese, selenium, and zinc (Sims and Wolf 1994), and some of these elements can be of concern for plant growth and even for human consumption if large quantities are added to soils. In a four-year study using broiler litter for several crops, only the soil concentrations of Zn and the nonmetal P were considered high enough to pose future problems for crops. Yields of canola, cotton, millet, and wheat increased, but those of peanuts were reduced by the applications of poultry litter at rates of 4.5, 9.0, and 13.5 Mg·ha^{-1}. Peanuts are sensitive to large concentrations of Zn, but whether yield reductions were due to excess Zn was not established in this study (Gasho et al. 2001).

The best control with such materials is to analyze them and limit their use to avoid undue concentrations in the soil. Some of these materials with borderline contents can be used for nonfood sources such as nurseries or golf courses. To avoid negative plant responses with such materials, it is desirable to keep the soil pH values above 7.0 to limit availability of the heavy metals.

Excess Salts

At times, adverse effects of large OM additions are due to excess salts. Large quantities of manure or biosolids may supply excesses of elements so that the salt content of the soil solution is elevated, thereby reducing the ability of the plant to take up water or nutrients. The problem is worse for salt-sensitive crops. Germinating seeds of many crops are particularly sensitive to excess salts. Fresh manure can be more of a problem than materials that have undergone some storage or composting. Liquid manures also can cause problems when applied to existing pastures unless they are diluted about fivefold.

The problem of excess salts from such materials can be alleviated by reducing the amount of OM applied, growing less sensitive crops, or by delaying planting until salts from the organic materials have been reduced by rainfall or irrigation.

Nutrient Imbalances

Organic matter additions may not provide balanced nutrition suitable for the succeeding crop, and may require some additions of fertilizers to provide the elements that may be in short supply. The imbalance is worsened by large additions of unbalanced organic matter. Some animal manure (bat guano, poultry solids without litter) and biosolids can be quite high in N and supply relatively small quantities of K. On the other hand, poultry litter and compost tend to supply relatively large amounts of P as compared to the K needs of many plants. Some leguminous residues, particularly hays, can supply disproportionately large amounts of N as compared to P and K. Corrections can be made by analyzing the OM and making adjustments as needed.

Pests

The presence of pests in OM can have a harmful effect on the succeeding crop. Fungi and bacteria present on plant residues may aggravate a number of plant diseases. Insects, nematodes, and weed seeds can be brought in with various crop residues or manures, seriously limiting the growth of the primary crop by their active competition or actual destruction of plant parts. For example, *Rhizoctonia* limb rot of peanuts increased with applications of poultry litter and probably contributed to a reduction in crops yields (Gasho et al. 2001).

The harmful effects of most pests introduced with OM can be eliminated or greatly reduced by composting the OM. The various pests and other methods of reducing their harmful effects are considered more fully in Chapter 5.

Chapter 3

Organic Matter As a Source of Nutrients

Organic matter can be a good source of nutrients for crops. Nearly all elements in the atmosphere and in the soil are taken up by plants. These elements are released when plants, and the humus that is formed primarily from plants, undergo decomposition. Sixteen of these elements are considered essential for all green plants. These elements and their chemical symbols are carbon (C), hydrogen (H), oxygen (O), nitrogen (N), phosphorus (P), potassium (K), calcium (Ca), magnesium (Mg), sulfur (S), boron (B), chlorine (Cl), copper (Cu), iron (Fe), manganese (Mn), molybdenum (Mo), and zinc (Zn).

Other elements may be essential or at least beneficial for some plants, but are not required by all green plants. The list includes aluminum (Al), cobalt (Co), chromium (Cr), nickel (Ni), selenium (Se), silicon (Si), sodium (Na), strontium (Se), titanium (Ti), and vanadium (V). Several of these elements (Al, Cr, Ni, Na, Se, Ti) more often have been associated with toxic effects than with stimulation of crops, but note these effects:

1. Al is necessary for the blue color of hydrangea, mitigates the toxic effects of Cu, and may increase availability of Fe for acid-loving plants.
2. Co is essential for N-fixing bacteria and the association of *Rhizobium* with legumes.
3. Na can substitute for some of the K required by certain crops.
4. Ni is beneficial for plants receiving most of their N from urea.
5. Se can replace S in some plants.
6. Si increases yield and strengthens the stems of rice, sugarcane, small grains, and some grasses while reducing susceptibility of these crops, tomatoes, and cucumbers to several fungal diseases.

7. Sr can be beneficial when Ca supplies are low.
8. Ti stimulates several crops, probably by its effect on chlorophyll.
9. Va seems to stimulate certain plants, probably by substituting for Mo in N fixation.

The C, H, and O are derived from air and water. Although air is the source of N used by plants, air N, except for the small amount converted to nitrates during electrical storms, cannot be used until it is transformed synthetically or by soil microorganisms. The synthetic process combines N with hydrogen to form ammonia (NH_3). The fixation of N by soil microorganisms is made possible by nonsymbiotic bacteria, primarily *Azotobacter*, or by those that are symbiotic with legumes, primarily *Rhizobium*. Soil is the primary source of N for nonlegumes. Much of the N used by plants is derived from SOM, various sources of OM, and fertilizers, with small amounts coming from water. But even with these sources, N originally comes from the atmosphere.

The other elements are derived primarily from soil minerals, OM, or fertilizers applied to the soil or plant, but water can supply considerable quantities of Na and Cl. Elements derived from the soil may be the result of soil weathering, microbial fixation, decomposition of organic matter, and/or the addition of various soil amendments.

ORGANIC MATTER AND NUTRIENT SUPPLY

Both OM and SOM are important sources of nutrients for crops. The amount of nutrients derived from organic matter available for crops is affected by the amounts present in OM and SOM, but as was seen in Chapter 2, the rate of release of the nutrients is dependent on the nature of the organic matter and such abiotic properties as O_2, moisture, and temperature levels of the soil.

Most of the nutrients supplied by organic matter are derived from the mineralization of organic matter as it decomposes. All nutrients present in OM and SOM become available for plants as organic matter decomposes, although some of it may be temporarily taken up by the microorganisms that cause the decay, while other nutrients may be subject to loss by leaching, and some N may also be lost by denitrification as it is converted to gaseous forms.

The majority of the elements (over 90 percent) in organic matter consists of O, C, and H (Table 3.1), which is present as various organic compounds. N, Si, and K make up about half of the remaining elements.

Nutrients supplied by various organic materials were orginally taken up by plants directly from the soil or indirectly through fixation from the air. These nutrients are potentially available for crops as organic matter decomposes, but until it does, the nutrients contained in OM are resistant to loss by leaching, immobilization, or denitrification. Once organic matter decomposes, the nutrients that it contained are subject to the same losses of leaching, immobilization, and denitrification as inorganic sources of nutrients. The delay in release of nutrients until decomposition takes place is usually beneficial for crops as fewer nutrients are lost but at times is detrimental if decomposition fails to keep up with the demands of the crop.

TABLE 3.1. Average elemental composition of corn plants*

Element	Composition (% dry weight)
Oxygen	44.43
Carbon	43.57
Hydrogen	6.24
Nitrogen	1.46
Silicon	1.17
Potassium	0.92
Calcium	0.23
Phosphorus	0.20
Magnesium	0.18
Sulfur	0.17
Chlorine	0.14
Aluminum	0.11
Iron	0.08
Manganese	0.04

Source: Miller, E. C. 1938. Plant Physiology, Second Edition. New York: McGraw-Hill Book Company. Reproduced by permission of The McGraw-Hill Companies.
*Dry weight of stems, leaves, cob, grain, and roots

NUTRIENTS RELEASED FROM ORGANIC MATTER

All nutrients that are absorbed by plants are potentially available for other plants as OM and SOM decompose. Some elements (C and O) are lost to the atmosphere as CO_2, and some N may also be lost as ammonia (NH_3), as oxides of N (NO or NO_2), or eventually as nitrogen (N_2). Much of the other elements and the N not volatilized become a part of the soil solution, or a part of SOM. The nutrients from the soil solution may be taken up by other plants or by various microorganisms as they break down the OM, held as cations by CEC, fixed by the soil, or lost by leaching. In time, microorganisms break down SOM, releasing the elements it contains, and microorganisms themselves or organisms that consume the microorganisms decompose to release elements they have accumulated, which again become part of the soil solution to repeat the process. The nutrients in organic matter become a part of a continuing cycle that nourishes plants, providing that enough new OM is added to maintain the continuity of the cycle.

The amounts of nutrients in organic matter that can become available for plants vary a great deal depending on the plant material, its age, the percentage of leaf tissue, the fertility of the soil, and the climate in which it was grown. Organic materials consisting of whole plants tend to have more N, P, K, Ca, Mg, S, and micronutrients than plants without leaves. The N contents of legumes usually are appreciably higher than nonlegumes, making these plants very attractive as sources of OM, since the amounts of N supplied by some of these crops often can be sufficient for the succeeding crop. The extent of nutrient variability is illustrated by the ranges of several nutrients that are taken up by some crops, given in Table 3.2.

The variability increases if animal products are included. Depending on the source, N in all organic materials may vary from about 0.25-10 percent; P_2O_5 from about 0.1-35 percent; K_2O from about 0.4-25 percent (equivalent to 0.25-10 percent N, 0.11-4.36 percent P, 0.08-29.05 percent K). The high values are associated with animal by-products. With the exception of a few high values for P_2O_5 (cotton seed meal, cotton hull ashes) and for K_2O (cotton hull ashes, beet sugar residue, distillery waste, and seaweed kelp), most plant materials analyze less than 5 percent N, 2 percent P_2O_5, and 5 percent K_2O (equivalent to 5 percent N, 0.87 percent P, 4.15 percent K).

TABLE 3.2. Variability of several nutrient elements in crop plants

Element	Range of composition (%)	Range in content (lb/acre*)
Nitrogen	1.5-6.0	50-500
Phosphorus	0.15-1.0	15-75
Potassium	1.0-5.0	50-500
Calcium	0.2-3.0	10-75
Magnesium	0.15-1.0	10-150
Sulfur	0.15-0.5	10-75

Source: Many different sources, including some of the senior author's analyses.

The average amounts of potentially available nutrients derived from plants are given in Table 3.3; from manure, sludge, and composts in Table 3.4; and from animal and plant by-products in Table 3.5.

Other Nutrients

Only N, P, and K contents are cited in Tables 3.3 through 3.5, but both OM and SOM are also good sources of the other essential elements. The amounts of Ca and Mg in SOM are lower than in most organic materials, but may be significant for crop production. The relatively low quantities of micronutrients (B, Cu, Fe, Mn, Mo, and Zn) in both OM and SOM usually are significant nevertheless because only small amounts are necessary for crop growth. The chelation effect of organic materials by which certain micronutrients are kept available enhance their significance.

Chelation of Micronutrients

Organic matter produces a number of chelates, substances that keep several metallic nutrient elements available over wide ranges of pH. Chelation of several micronutrients (Fe, Cu, Zn) is important in many parts of the world, because of the tendency of these elements to become unavailable in high-pH soils or in the presence of large amounts of heavy metals.

TABLE 3.3. The composition of some common organic materials

Material	Moisture (%)	Approximate pounds per ton of material at the indicated moisture content		
		N	P_2O_5*	K_2O**
Alfalfa hay	10	50	11	50
Alfalfa straw	7	28	7	36
Barley hay	9	23	11	33
Barley straw	10	12	5	32
Bean straw	11	20	6	25
Beggarweed hay	9	50	12	56
Bermudagrass hay				
Coastal	10	50	8	40
Common	10	32	9	43
Bluegrass hay	10	35	12	35
Buckwheat straw	11	14	2	48
Clover hay				
Alyce	11	35	–	–
Bur	8	60	21	70
Crimson	11	45	11	67
Ladino	12	60	13	67
Subterranean	10	70	20	56
Sweet clover	8	60	12	38
Corn stover, field	10	22	8	32
Corn stover, sweet	12	30	8	24
Cowpea hay	10	60	13	36
Cowpea straw	9	20	5	38
Crested wheatgrass	10	62	11	44
Fescue hay	10	42	14	47
Fieldpea hay	11	28	11	33
Fieldpea straw	10	20	5	26
Horse bean hay	9	43	–	–
Lezpedeza hay	11	41	8	22
Lezpedeza straw	10	21	–	–

Material	Moisture (%)	Approximate pounds per ton of material at the indicated moisture content		
		N	P_2O_5*	K_2O**
Oat hay	12	26	9	20
Oat straw	10	13	5	33
Orchard grass hay	10	45	14	55
Peat, muck	30	45	9	15
Peat, sphagnum	–	11	2	–
Peanut vines	10	60	12	32
Rice straw	10	12	4	27
Ryegrass hay	11	26	11	25
Rye hay	9	21	8	25
Rye straw	7	11	4	22
Sorghum stover	13	18	4	37
Soybean hay	12	46	11	20
Soybean straw	11	13	6	15
Sudangrass hay	11	28	12	31
Velvet bean hay	7	50	11	53
Vetch hay, common	11	43	15	53
Vetch hay, hairy	12	62	15	47
Wheat hay	10	20	8	35
Wheat straw	8	12	3	19

Source: Modified by authors' data from Lorenz, O. A. and D. N. Maynard. 1988. *Knott's Handbook for Vegetable Growers,* Third Edition. New York: John Wiley and Sons.
*Divide by 2.2914 to obtain P.
**Divide by 1.2046 to obtain K.

Within the plant, common chelated metal atoms are associated with the heme group as iron porphyrin or as chlorophyll. Malic and oxalic acids, supplied by OM, are recognized as having chelating properties. Root exudates capable of forming iron complexes that stay available for long periods are very useful in this respect. The plant chemical (α)-ketogluconic acid has been designated as being highly useful for this purpose (Webley and Duff 1965), but there are

TABLE 3.4. Common macronutrient contents of manures, composts, and biosolids

Material	N %	N lb/ton	P_2O_5** %	P_2O_5** lb/ton	K_2O*** %	K_2O*** lb/ton
Bat guano	7.5	150	3.0	60	1.5	30
Beef, feedlot	0.71	14	0.64	13	0.9	18
Compost*	0.9	18	2.3	46	0.1	4
Dairy	0.56	11	0.23	5	0.6	12
Dairy, liquid	0.25	5	0.05	1	0.25	5
Duck	1.1	22	1.45	29	0.5	10
Goose	1.1	22	0.55	11	0.5	10
Horse	0.7	14	0.25	5	0.7	14
Poultry, no litter	1.55	31	0.9	18	0.4	8
Poultry, liquid	0.15	3	0.05	1	0.3	6
Sewage sludge	0.9	18	0.8	16	0.25	5
Sewage sludge, act.	6.0	120	2.5	50	0.2	4
Sheep	1.4	28	0.5	10	1.2	24
Swine	0.5	10	0.32	6	0.45	9
Swine, liquid	0.1	2	0.05	1	0.1	2

Source: Many different sources, including some of the authors' analyses.
*Finished and screened aerated sludge compost
**Divide by 2.2914 to obtain P.
***Divide by 1.2046 to obtain K.

probably many more chelating compounds in plants or formed as organic materials decompose.

The addition of farmyard manure to high-pH soils for the purpose of providing chelates to correct lime-induced chlorosis, usually due to lack of available Fe at the high pH, however, has not been very useful. The failure of manure to correct the problem is attributed to formation of considerable carbon dioxide (CO_2) as this organic materials decomposes. The CO_2 favors the formation of bicarbonates, which reduce Fe uptake and translocation within the plant.

TABLE 3.5. Major nutrient contents of plant and animal by-products (lb/ton)

Product	N	P_2O_5*	K_2O**
Animal tankage	140	180	–
Ashes, cotton hull	0	80-140	22-30
Ashes, hardwood	0	20	100
Beet sugar residue	60-80	0	160-200
Bone meal	90	144	–
Bone, precipitated	–	700-920	–
Bone tankage	60-200	140-400	60-180
Castor pomace	100-120	40	20
Cocoa shell meal	60	40	20
Cocoa tankage	80	20	40
Cottonseed meal	120-180	40-60	20-40
Crab scrap	60	60	0
Distillery waste	20	10	280
Dried blood	160-280	10-30	10-16
Dried king crab	180-240	–	–
Fish, acid	120	120	–
Fish meal	100-200	100-260	–
Fish scraps, fresh	40-160	40-120	–
Garbage tankage	60	40	60
Hoof and horn meal	260	–	–
Linseed meal	100	20	20
Olive pomace	20	20	10
Peanut hulls	30	4	20
Rapeseed meal	100-120	–	–
Seaweed kelp	40	20	80-260
Shrimp scrap, dried	140	80	–
Soybean meal	120	20	40

TABLE 3.5 (continued)

Product	N	P$_2$O$_5$*	K$_2$O**
Steamed bone meal	60	400	–
Tobacco dust and stems	–	40	100
Winery pomace	30	30	20
Wool wastes	140	–	–

Source: Collings, G. H. 1947. *Commercial Fertilizers*, Fifth Edition. New York: McGraw-Hill Book Company, Inc.; Follett, R. H., L. S. Murphy, and R. L. Donahue. 1981. *Fertilizers and Soil Amendments*. Englewood Cliffs, NJ: Prentice-Hall Inc.; Lorenz, O. A. and D. N. Maynard. 1988. *Knott's Handbook for Vegetable Crops*, Third Edition. New York: John Wiley and Sons.
*Divide by 2.2914 to obtain P.
**Divide by 1.2046 to obtain K.

NUTRIENT RELEASE

The actual amounts of nutrients released from OM and SOM for a crop may be only a small fraction of the total present. At times, there may be no release from OM additions and there actually can be a temporary reduction of plant-available N. As mentioned earlier, the amounts released from both OM and SOM depend on climatic factors of temperature, moisture, and soil oxygen content, but the release of N and possibly P and S from OM also depends on the ratio of these elements to C in the material and at times to contents of polyphenols or lignin.

The content of nutrients present in SOM is relatively uniform in all soils, but that in OM can be quite variable among plant and animal sources. Although usually having more N, P, and S than many OM materials, SOM is usually lower in K, Ca, Mg, and micronutrients. The composition of SOM is approximately 5 percent N, 0.5 percent P, and 0.5 percent S on a dry weight basis, with smaller but significant amounts of K, Ca, Mg, and micronutrients. Each percent of SOM in mineral soils contains about 1,000 lb N, 100 lb P, and 100 lb S per plowed acre.

Many animal by-products have high N values. For a long period, many of these were used as sources of N for a number of plants. Their high N contents and the resistance of the N to leaching made them excellent N sources for a number of crops. They probably would still be used for such purposes, but the prices of these materials have risen because of their value as pet foods, making them noncompetitive with synthetic sources of N.

Release of N, P, and S

Most nutrients in OM and SOM are present as organic molecules and are not available to plants until they are converted to simpler inorganic forms. Most of the N, which is present in SOM at about 5 percent of its weight, is present as a constituent of proteins. Plants cannot use protein-N or N in most other organic compounds until the N is converted to ammonium (NH_4) or nitrate (NO_3) forms. The conversion takes place in a series of steps, with NH_4-N being formed first and then converted to NO_3-N by other microorganisms. The conversion is dependent on a suitable soil environment, with one of the more important criteria being the presence of sufficient O_2. As we shall see, OM and SOM affect pore space, which is largely responsible for the amount of O_2 available for plants and microorganisms.

The release of N, P, and S from OM is closely related to the ratio of these elements to C in the added material. A narrow ratio of these elements to C usually allows fast nutrient release, but a wide ratio may require additional inputs of these elements as fertilizers to avoid reductions in crop yields. The reason for poor performance with the wide ratio is that the supply of the nutrients in the decomposing organic matter is inadequate to meet the microorganisms' needs, causing them to remove the missing elements from the soil solution, thereby lowering the amounts available for the crop.

As mentioned earlier, the ratio of C:N in SOM is about 10:1, while that of C:P and C:S is close to 100:1. There is a very fast release of these elements if the ratio of N in the OM is not much wider than 10:1 and of P and S if the ratio of C to these elements is not much wider than about 100:1.

The speed of release is usually reduced as the ratio of N, P, or S to C widens because the decomposition of OM may have to wait until there are sufficient quantities of N, P, or S to meet the requirements of the microorganisms that decompose the OM. If ratios of N, P, and/or S to C in OM are appreciably wider than those in SOM, decomposition may still proceed normally if the surrounding environment contains enough of these elements. The microorganisms then will use the available soil N, P, or S to supplement that which is missing in the OM. In so doing, the microorganisms often can remove enough N, P, and/or S to affect the growth of higher plants. In time, these elements will be released as the microorganisms themselves undergo decomposition.

The release of N at different ratios to C has been studied more widely than the release of P or S for several reasons. Shortages of N following incorporation of OM with wide C:N ratios are much more common than shortages of P or S following incorporation of materials with either a wide C:P or C:S ratio. The C:N ratio is the more important of the three ratios because of the amounts of N needed, and the importance of organic materials as a source of N for most crops. Both P and S are also present in the soil in inorganic forms and can be readily supplied by additions of inexpensive chemical fertilizers that do not leach as readily as the inexpensive sources of N.

The C:N ratios of a number of materials that may be added to soils are listed in Table 2.1. Although the release of N is closely related to these ratios, exceptions need to be noted. Release of N from OM can still be expected if the C:N ratio is somewhat wider than the average ratio of 10:1 found in SOM. Obviously there would be no problems with the 12:1 found in some sewage sludges (biosolids) or the about 12:1 ratio found in young legume materials. Nor does there appear to be any problem with N release from older legumes with a C:N ratio of less than 15:1. There are problems with some legumes having C:N ratios from 15:1 to 35:1, but not with sheep or chicken manure with C:N ratios of about 17:1, nor even with cattle manure with a C:N ratio of about 30:1. The relatively poor release of N from some leguminous materials with relatively narrow ratios of 15:1 to 35:1 appears to be due to the complexing of proteins (N source) by polyphenols, which protects the proteins from decomposition to a certain extent. On the other hand, some woody materials with very wide C:N ratios of 75:1 to 100:1 may decompose quite rapidly because their lignin contents, which normally tend to protect cell sugars from rapid decomposition, are relatively low.

There appears to be a slowdown in decomposition of OM as the C:N ratio widens beyond 20:1, with nitrogen deficiencies of the succeeding crop appearing as the ratio widens appreciably beyond 30:1. The presence of N deficiencies is avoided if sufficient N is available in the root zone from past accumulations or fertilizer additions or if sufficient time is allowed between incorporating OM and the planting of the crop. The immobilization of N by microorganisms is of shorter duration if the rate of decomposition is hastened by elevated soil temperatures, ample moisture, and O_2, and by shredding the OM into small pieces. Often, there is insufficient time to allow for sufficient decomposition before planting the next crop, making it very important to add the extra N early enough to avoid problems.

Release of Nutrients Other Than N, P, or S

The release of K, Ca, Mg, and micronutrients (B, Cl, Cu, Fe, Mn, Mo, and Zn) from organic materials does not appear to be related to any specific ratios of these elements to C. Nevertheless, the release of some of the Ca, Mg, and Cu may be slowed as wide ratios of C:N reduce the rate of decomposition, because to some degree they become a part of the organic matter or are tightly bound to it. The release of K, Cl, and the other micronutrients are much less affected since a good portion of these elements remain uncombined or so weakly held in the plant that they are quickly released soon after OM is incorporated.

INDIRECT EFFECTS OF ORGANIC MATTER ON NUTRIENT AVAILABILITY

In addition to the direct effect of organic matter contributing nutrients for plants by assimilation and later release of nutrients, organic matter indirectly affects the amount of nutrients available to plants by (1) altering several soil properties that affect the availability of several nutrients; (2) helping to create a balanced ecological system favoring several biological processes that increase available N and P and also foster the decomposition of organic matter, releasing large quantities of valuable nutrients; and (3) providing the energy that makes possible processes such as the fixation of atmospheric N to forms available to plants.

Porosity, cation exchange capacity, buffering ability, and ecological balance are the major soil properties affecting nutrient availability that are changed by organic matter. Although these changes in soils brought about by organic matter are usually beneficial, they can at times lessen the amount of nutrients available for the crop by increasing leaching, immobilization, and denitrification.

Changes Increasing Nutrient Availability

Porosity

Organic matter tends to improve soil porosity, allowing for a better balance between soil air and moisture. The balance has a bearing on the amounts of N fixed by both symbiotic and nonsymbiotic fixation and the amounts of all nutrients made available from the decomposition of organic matter because most microorganisms involved in the processes need both moisture and O_2 for maximum efficiency. The balance between O_2 and moisture is also essential for the uptake of nutrients and water since a lack of O_2, which can occur with excess water, greatly inhibits uptake of nutrients and water. More about organic matter and porosity is presented in Chapter 4, which deals with the physical effects of organic matter.

Cation Exchange Capacity

Soil organic matter and clay are the sources of soil CEC, by which the cations are held in an available form and are readily exchanged for other cations. The exchange quite often is made as a hydrogen ion (H^+), from plant roots, replaces one of the cations in the exchange complex. By reducing fixation and leaching losses of the cations, the CEC helps maintain a more constant supply of these nutrients, thereby potentially increasing crop yields.

The amount of CEC contributed by SOM varies from about 30 to 600 meq/100 g with common values of 150-250 meq/100 g of dry material (Cooke 1967). (In more modern terms, these values equal 30 to 600 cmoleckg^{-1} and 150-250 cmoleckg^{-1}). In sands or soils dominated by 1:1 clay minerals, the major portion of the CEC is derived from SOM. Per unit weight, SOM has much greater CEC than most clay minerals. Its CEC is so high that CEC of many soils, particularly sandy ones, can be substantially increased by growing sods and hays or by adding substantial annual dressings of OM such as animal manure.

Buffering Effects

The buffering effect of soils, which resist soil pH changes, increases with increasing contents of clay and organic matter. Humus or SOM tends to be acidic when free of electrolytes and is readily dispersed or dissolved. Ca or Mg, commonly present in large amounts in alkaline soils, forms humates that are not readily dissolved or dispersed, thus restricting pH changes on the alkaline side. Aluminum and Fe, usually present in acid soils, also tend to form humates that are not readily dispersed or solubilized, thus restricting pH changes in the acid range.

The buffering effect has significant impact for agriculture. While more liming materials are needed to raise the pH of a soil rich in organic matter, the effects are much longer lasting. A stable satisfactory pH is a substantial asset, as it allows for uniform growth of most plants, despite the additions of acid-forming materials from fertilizers or the deposition of acid rain.

By limiting pH change in acid soils, buffering greatly reduces the chances of problems from excess Al, Fe, and Mn. As values fall below about pH 6.0, increasing amounts of these elements are brought into solution. Aluminum in relatively small quantities can depress growth of a number of plants, partly because of its ability to immobilize P. Organic matter, because of its buffering capacity, greatly reduces the immobilization of P by Al, Fe, and Mn. Although less common, excess Fe and Mn, also increased with lower pH values, can restrict plant growth.

Ecological Balance

Ecological systems affect the kind and amounts of organisms active in the soil. Organic matter, by affecting porosity, limiting marked changes in soil temperatures, helping to maintain adequate soil moisture levels, and supplying needed energy for many vital soil reactions, affects the kind, numbers, and activity of most microorganisms.

The changes brought about by organic matter help maintain a balanced system of soil flora and fauna. The balanced system helps to increase available nutrients through symbiotic relationships, decomposition of organic matter, and restriction of organisms that can injure crops. Some of the nutrient effects of microbiological systems are considered here, but the biological effects are covered more fully in Chapter 5.

Nitrogen Fixation

Nitrogen fixation can be defined as the conversion of air N from a form unavailable to plants to one that is readily absorbed. It is the primary mechanism by which atmospheric N initially enters the plant-soil environment. Nitrogen fixation occurs by the action of symbiotic and free-living organisms. Symbiotic N fixation takes place as microorganisms, primarily *Rhizobium* spp. living mostly in the nodules on roots of legumes, are able to convert the nonavailable air N into plant-available forms. The nonsymbiotic fixation from air is accomplished by free-living organisms. In most soils, this kind of N fixation is accomplished primarily by bacteria (predominantly *Azotobacter* spp.). In the wet soils used for rice production, free-living blue-green algae fix large quantities of N.

The mineral elements Ca, K and Mg, B and Mo are closely involved in symbiotic N fixation. The Ca, K, and Mg are held in an available form in the exchange complex and, as has been pointed out, SOM can be an important part of CEC. A large CEC sufficiently saturated with Ca, Mg, and K can help ensure sufficient supplies of these essential cations for the maintenance of the host and development of the symbiotic microorganisms.

Although a good supply of most nutrient elements tends to increase N fixation, that of N usually decreases it. An ample supply of available N (NH_4-N and/or NO_3-N), whether from fertilizer, manure, sludge, compost, or large amounts of decomposing OM, will reduce the amounts of N fixed by the symbiotic process as the host plant utilizes the readily available forms.

The importance of organic matter as an energy source for the symbiotic and nonsymbiotic fixation of atmospheric N normally unavailable for plants to forms that are readily available has already been touched upon. The amounts of N fixed by the symbiosis process with several legumes are presented in Table 3.6. The amounts fixed by nonsymbiotic organisms *(Azotobacter, Azospirillum, Anabena, Clostridiumi, Cyanobacter, Frankia)* in many soils are much smaller, probably only about 6 lb per acre, but blue-green algae in the wet conditions of rice culture fix about 70 lb N per acre, although laboratory experiments with these organisms indicate a potential fixation of 500 lb per acre per year (Cooke 1967).

TABLE 3.6. Approximate amounts of nitrogen fixed by various legumes

Crop	lb/acre*
Alfalfa	190
Beans	40
Clover	
Crimson	160
Ladino	175
Red	115
Sweet	120
White	100
Cowpea	90
Hairy vetch	165
Kudzu	110
Lespedesza (annual)	85
Pea (English)	80
Peanut	50
Perennial peanut	210
Soybean	110
Winterpea	60

Source: Follett, R. H., L. S. Murphy, and R. L. Donahue. 1981. *Fertilizers and Soil Amendments.* Englewood Cliffs, NJ: Prentice-Hall Inc.; Pieters, A. J. and R. McKee. 1938. The use of cover and green-manure crops. In *Soils and Men: Yearbook of Agriculture.* Washington, DC: U.S. Government Printing Office; Tisdale, S. L. and W. L. Nelson. 1975. *Soil Fertility and Fertilizers,* Third Edition. New York: Macmillan.
*Multiply by 1.12 to obtain kg/ha.

Mycorrhizae and Available Phosphorus

Mycorrhizae are fungi that live in a symbiotic relationship with a number of plants. They benefit plants in several ways, which are presented in greater detail in Chapter 5, but their effects on nutrients need to be addressed here.

Typical kinds of mycorrhizae are established with dominant plants of an ecological system. Basically, there are two types of mycorrhizae. Ectomycorrhizae tend to form associations with many trees and shrubs in temperate zones. They invade plant roots but remain outside the cortex cell walls, never penetrating the cell. The other type of mycorrhiza, the endomycorrhiza, also called the arbuscular mycorrhiza or AM, is much more widespread, extending over diverse climates and forming associations with most agronomic crops, vegetables, fruit crops, and a number of forest and tree crops. It penetrates the root cell walls, where it forms structures within the plant that help to exchange nutrients with the host plant or store sugars for later use.

Both types influence nutrient availability for plants in the following ways:

1. The fungal hyphae greatly extend the capacity and effectiveness of the roots, essentially increasing the root system at least severalfold.
2. The hyphae, by extending the root system, make it possible to tap into additional sources of nutrients and more efficiently utilize low levels of nutrients or nutrients that are not in solution. They add to the availability and/or uptake of several elements, but that of P is particularly notable.
3. The hyphae are able to reach into smaller pores than root hairs, making some water and nutrients available to plants that would not otherwise be available.
4. The mycorrhizae secrete hormones that stimulate plants, which may be responsible for improved germination of vegetable seeds.
5. Mycorrhizae inhibit the uptake of excess salts and toxic metals and are beneficial in increasing the host's tolerance to other stresses, such as drought conditions, low fertility, and temperature extremes, all of which can affect the uptake and utilization of several nutrients.

In low-fertility soils, mycorrhizae are probably responsible for the survival of many plants. They can tap into sources of P several centimeters away from plant roots, increasing the amount of P potentially available to plants. By absorbing P ions as soon as they go into solution and transporting them into the plant, mycorrhizae keep a great deal of P available that might otherwise quickly convert to unavailable forms in the soil.

Mycorrhizae are increased in soils with high organic matter but may be largely eliminated by soil sterilization and application of some pesticides or large amounts of fertilizer. Steam sterilization of most soils can eliminate enough mycorrhizae so that plant growth is severely inhibited unless there is an immediate application of soluble P.

Although ectomycorrhizae are symbionts (live in a symbiotic relationship with the host plant), they can survive independently, allowing them to be reproduced in cultures and making it possible to inoculate plants with them to reduce certain stress factors, although the ability to reduce stress varies with different strains of ectomycorrhiza.

Ectomycorrhiza inoculations of tree seedlings have been successfully used for planting in problem sites, such as toxic coal or mine spoils. Soils in these areas usually have very low pH value (<4.0) and may contain excess metals, making them almost unsuitable for planting Virginia or red pine seedlings. Inoculation with the ectomycorrhiza *Pisolithus tinctoris* provided a better stand and growth of pines on a Kentucky soil with pH 3.0 than if the pines were inoculated with *Thelephora terrate* mycorrhiza. Some natural ectomycorrhiza benefit willow and poplar grown on iron tailings but have little or no benefit for nurturing growth on copper tailings (Maronek et al. 1981).

Changes Accentuating Losses of Nutrients

While organic matter tends to greatly increase the amounts of nutrients available for crops, it may also be responsible for nutrient loss. Changes brought about by organic matter that adversely affect amounts of nutrients by increasing losses due to denitrification, leaching, and immobilization are briefly presented here.

Denitrification

Nutrient loss may be accentuated by organic matter because it tends to increase wet conditions, thus increasing the loss of N by denitrification, whereby N is released to the atmosphere. Organic matter left on the surface as mulch can potentially affect considerable N lost by denitrification, partly due to the wet conditions maintained at the soil surface. Ammoniacal compounds and urea are particularly prone to denitrification losses, and direct application of these fertilizer sources to organic matter, as is done with dribbling (fertilizer solutions dropped as a stream), should be avoided. Incorporating these fertilizer materials in the soil tends to reduce losses.

Leaching

Losses of soluble nutrients (primarily N, but also K and Mg in low-CEC soils) tend to increase as SOM is increased because of the improved porosity with the higher levels of organic matter and also because of much higher moisture levels maintained in the surface soil, particularly if the OM is retained as mulch. Losses can be reduced by substituting less soluble or coated forms of fertilizers or by reserving some of the N for later side-dressings.

ACCOUNTING FOR NUTRIENTS IN ORGANIC MATTER

The nutrients contained in SOM or added OM such as crop residues, manure, compost, biosolids, or other waste materials logically should be taken into consideration when planning a fertilizer program. But there are several problems with basing fertilizer recommendations on the amount of added OM, the presence of SOM, or nutrients indirectly gained or lost because of organic matter.

One of the problems in accurately adjusting fertilizer recommendations to compensate for nutrient changes associated with organic matter is the difficulty of accurately measuring nutrients added by OM. Average values often are not very accurate as there is considerable variability in composition. It is possible to get a more accurate estimate of elements added as OM by obtaining an analysis of the material and multiplying it by amounts applied, but obtaining these values can be time consuming and expensive. Quantifying the values for cover crops and crop residues is more difficult than obtaining values for some organic materials, such as manure and composts. It is possible to harvest several square yards of material and obtain an analysis, but this usually involves considerable time and expense and may not reflect true values as it is difficult to obtain a value for nutrients contributed by roots.

The speed at which these materials decompose and the nutrients become available for the primary crop also can add to the difficulty of compensating for the nutrients supplied directly from the decomposition of OM and SOM or indirectly, as these materials affect several soil properties. The rate of release can be quite variable because, as has been noted before, the decomposition rate is dependent on such variables as C:N ratio, biphenol and lignin content of the organic material and moisture, temperature, and O_2 content of the soil.

Deciding whether extra nutrients need to be added as organic materials are added to the soil may pose a problem. Succulent plants usually do not need extra nutrients for decomposition because these materials, which are rich in sugars and proteins, decompose very rapidly, releasing nutrients quickly. On the other hand, grasses and other nonlegumes tend to decompose much more slowly, especially as these materials approach maturity. The C:N ratio of most plant materials widens as the plant ages or if it no longer has leaves, slowing down its decomposition and nutrient release. Determining whether the C:N ratio is wide enough to cause problems for the succeeding crop may be costly or difficult, since most agricultural laboratories are not equipped to determine C. They usually can measure total N and use that as a criterion to decide when to supply extra N when plants or other materials with wide C:N ratios are incorporated.

Although not infallible, basing N applications on the C:N ratio can be helpful and is often used when incorporating plant residues. Since most agricultural laboratories might not be able to provide quick C analysis but can provide analysis for N, the question whether the OM addition will release N, and therefore permit reduction in the amount of additional N needed, or whether the addition of OM will require more N, is often decided by using N analysis of mature residues.

No initial release of N can be expected if the N content of the incorporated material is less than about 1.75 percent, and N shortages can be expected if the N content is less than 1.5 percent. The addition of fertilizer N to take care of materials with N contents less than 1.5 percent still may not be needed if the soil has high levels of available N. A soil test of 25-50 parts of NH_4-N + NO_3-N per million parts of soil (ppm or mg per kg) will be sufficient to take care of the extra N needed to decompose most additions of crop residues. If available N (NH_4-N + NO_3-N) is appreciably lower than 25 ppm, the addition of about 50 pounds N/acre (56 kg/ha) usually will compensate for the low N in many crop residues.

From a practical standpoint, an estimate of the amount of N required if OM tests less than 1.75 percent N can be made, assuming that 30 lb of N are required for the microbial breakdown of 2000 lb of most organic materials. On this basis, a ton of rye straw testing 0.75 percent N supplies only 15 lb N per ton, requiring another 15 lb of N to prevent the addition of the straw from tying up N as it decomposes.

The addition of young crops, most legumes, manure, biosolids, and other materials with C:N ratios narrower than 30:1 can be expected to release N and other valuable nutrients. But the amounts released are difficult to quantify because the breakdown is subject to all the factors that affect activity of microbial populations. Further problems arise as the nutrients finally released from OM are subject to loss by leaching, volatilization, and immobilization by microorganisms or clay. The amounts of nutrients available for the crop following the addition of the OM may be only a small portion of what was originally present in the OM. The addition of animal manure is cited next as an example.

Nitrogen Release from Manure

Not all the elements in OM will be released in time for the succeeding crop. Even if C:N ratios are relatively narrow, as in the case of manure, only a portion of the elements can be expected to be released in the first year after incorporation. Usually, only about one-third to one-half of the N in manure will be available in the first crop year, with most of the remainder becoming available in the next year or two. Accordingly, in a 10-ton per acre application of manure, which contains 50-200 lb of N depending on the type of manure and how it was handled, often only 17-67 lb of the N will be available to the crop in the first year following the addition. The actual nutrient release depends on composition of the manure and microbial decomposition, the latter of which is influenced by factors listed earlier in this chapter.

Despite these limitations, there is a consensus that N recommendations for the crop following the addition of the organic matter need to be adjusted depending upon the previous crop or manure application. Some average values have been established over the years that can be very useful. The amounts suggested, however, vary depending on the type of crop and its management or yield goals.

Reductions in Fertilizer Recommendations Based on Manure Applications

Although amounts of nutrients released from manure vary considerably, some reductions in fertilizer for the newly planted crop usually can be made without harming yields.

Cooke (1967) estimates that an application of farmyard manure in the range of 12-15 tons carries about the same nutrients as 450 lb of a 12-5-12 mixed fertilizer or about 54 lb N, 22 lb P_2O_5, and 54 lb K_2O (equivalent to 54 lb N, 10 lb P, and 45 lb K). On this basis, a 10-ton manure application (a more common unit application) would supply about 40 lb N or K_2O and only 16 lb P_2O_5 (equivalent to 40 lb N, 7 lb P, and 33 lb K), and fertilizer recommendations for the succeeding crop could be lowered accordingly.

Most recommendations for fertilizer application following manure tend to consider only the N supplied by the manure. It is common practice to lower the N applied for the crop following the manure application, but the amount of reduction often depends on the crop grown, the yield goals, or type of management. For example, Michigan State University recommends a 40 lb reduction in applied N per acre if 10 tons of manure are applied for corn or potatoes, but only about 30 lb reduction for small grains. The reduction is increased to 50 lb for high-level management of pasture or hay but 60 lb for low-level management of pasture or hay (Table 3.7).

TABLE 3.7. Reduction in nitrogen recommendations for several crops grown in Michigan following a legume or manure application (lb/acre)

Previous crop or manure application	Barley, oats, rye	Wheat	Field beans or soybeans	Grass, pasture or hay		Potatoes (cwt/acre)		
				A*	B*	250-349	350-449	450-550
Legume + manure**	10	10	0	0	0	30	60	90
Good legume	10	10	10	0	0	70	100	130
Manure**	10	30	10	0	0	90	120	150
No legume, no manure	40	60	40	60	100	120	150	190

Source: Warnecke, D. D., D. R. Christenson, and R. E. Lucas. 1976. *Fertilizer Recommendations for Vegetables and Field Crops.* Extension Bulletin E-550. Farm Science Series. East Lansing, MI: Cooperative Extension Service, Michigan State University.
*A = Low maintenance; B = High maintenance.
**Manure application = 10 tons/acre.

Reductions in Fertilizer Recommendations Based on Previous Crop

Reductions in N applications are recommended when a legume crop precedes the primary crop, with greater reductions when both manure and a legume crop precede the primary crop (Tables 3.7, 3.8). As indicated in Table 3.7, an application of 10 tons of manure/acre warranted a reduction of 30 lb N/acre, while a reduction of 50-60 lb N/acre are suggested if a good legume precedes the potato crop and a reduction of 90-100 lb N/acre are suggested if both legume and manure are added. A similar typical example of variations in N recommendations made for corn grown in Michigan depend on whether an application of 10 tons/acre of manure, a good legume, or a combination of manure and legume precede the corn crop as given in Table 3.8. The amount of N recommended for each of these situations varies depending on yield goals.

Generally, appreciable reduction in recommended N is suggested if the previous crop was a legume rather than corn, small grains, or grass crops. For corn or grain sorghum, Purdue University, Indiana, recommends a greater reduction in N if the previous legume was "good," consisting of 5 plants of red clover or alfalfa per ft^2, than if it was only "average," consisting of 2-4 plants/ft^2. Reductions based on either of these categories were greater than if soybeans were the pre-

TABLE 3.8. Nitrogen recommendations for corn grown in Michigan based on previous legume crop and/or manure applications and yield goals

Previous crop or manure application	Yield goal (bu/acre)			
	60-89	90-119	120-149	150-180
	N (lb/acre)			
No legume, no manure	70	100	150	200
Manure 10 tons/acre	30	60	110	160
Good legume	10	40	90	140
Manure + good legume	0	0	50	100

Source: Warnecke, D. D., D. R. Christenson, and R. E. Lucas. 1976. *Fertilizer Recommendations for Vegetables and Field Crops.* Extension Bulletin E-550. Farm Science Series. East Lansing, MI: Cooperative Extension Service, Michigan State University.

vious crop. Suggested N reductions were also adjusted to yield goals (Table 3.9).

Adjustments in N applications for corn or wheat in Illinois also varied depending on the number of alfalfa or clover plants per square foot. Some reductions in N application were suggested for corn the second year following a good crop of alfalfa (Table 3.10).

Release of Nutrients from SOM

The release of N from SOM has been studied to a greater degree than the release of other elements because of the large amounts of N in SOM, the importance of N in plant nutrition, and the difficulty of maintaining adequate N for many crops. The release appears to be closely related to loss of SOM, but the quantity of nutrients released as SOM decomposes is difficult to measure because of contributions made by additions of OM, immobilization of N by microorganisms, and losses of N due to leaching and volatilization. These difficulties

TABLE 3.9. Adjusted nitrogen recommendations for corn or grain sorghum grown in Indiana based on previous crop and yield goals*

	Yield goal (bu/acre)				
	100-110	111-125	126-150	151-175	176-200
Previous crop	N (lbs/acre)				
Good legume (5 plants alfalfa or red clover/ft^2)	40	70	100	120	150
Average legume (2-4 plants alfalfa or red clover/ft^2)	60	100	140	160	180
Soybean-legume seeding of alfalfa or red clover	100	120	160	190	220
Corn, small grains, hay, pasture, etc.	120	140	170	200	230

Source: Plant and Soil Test Laboratory. Agronomy Department, Purdue University, Lafayette, IN. Courtesy of Professor R. K. Stivers.
*Some factors affecting yield goals are water availability and other climatic variability, potential of varieties used, inherent fertility of the soil, and/or potential prices of grain and fertilizer.

TABLE 3.10. Adjusted nitrogen recommendations for corn and wheat grown in Illinois based on previous crop

Previous crop	Plants/ft.^{2}	Nitrogen reduction (lb/acre)	
		Corn	Wheat
First year after alfalfa or clover	5	100	30
	2-4	50	10
	<2	0	0
Second year after alfalfa or clover	5	30	0
	<5	0	0
Soybeans		40	10

Source: Hoeft, R. G., T. R. Peck, and L. V. Boone. 1994. Soil testing and fertility. In *Illinois Agronomy Handbook 1995-1996,* p. 70-101. University of Illinois at Urbana-Champaign, College of Agriculture Cooperative Extension Service, Circular 1333.

probably explain the considerable variation in N recommendations based on SOM.

Some estimates of N release from SOM suggest values from about 20-40 lb/acre (22.4-44.8 kg/ha) per year for each percent OM, which is only 2-4 percent of the total N present, but even this small amount has been questioned as being too high, especially if factors favoring decomposition are not satisfactory.

As indicated earlier, the release of nutrients from SOM, primarily N, P, and S, depend on soil moisture, temperature, and oxygen. The soil pH and nutrient content will also have a bearing on nutrient release from SOM. Release of N, P, and S from SOM on most commercial farm operations having irrigation is seldom limited because of moisture, pH, or nutrient content, but is frequently adversely affected by temperature and soil O_2. Little if anything can be done to change adverse soil temperatures, except perhaps with mulch (see Chapter 8), but much can be done to improve soil O_2 levels, since adverse levels are often due to poor management. The improvement of soil O_2 has been covered in many treatises, including the senior author's *The Fertile Triangle* (Wolf 1999), but, essentially, corrective measures include the following:

1. Adding more OM
2. Reducing tillage

3. Avoiding working soil when wet
4. Limiting use of heavy machinery, especially when soils are wet
5. Reducing impact of overhead irrigation
6. Avoiding use of irrigation waters with either low or high salt contents
7. Maintaining a calcium saturation at about 70 percent of the CEC

The O_2 level may also vary depending upon the soil texture, with greater amounts usually present in sands, decreasing in order of loams, silt loams, and clays. The variable O_2 content alters the rate by which OM is broken down, and this is reflected in the amount of SOM present and the estimated release of nutrients in that which remains.

Some soil laboratories use the estimated release of N from SOM as a partial basis for recommending N for the cash crop, taking into consideration the variable release as influenced by soil class. For example, A & L Laboratories estimate the release of N from SOM to be in the order of sandy loams → silt loams → clay loams (Table 3.11).

The adjustment of fertilizer recommendations for different soil textures is only partly justified by the differences in SOM. Variation in N leaching according to soil texture is also an important factor. For example, N recommendations for small grains grown in Indiana, which vary with the soil texture, probably take into consideration the variability of N leaching in different soils as well as the variability in O_2 (Table 3.12).

Recommendations for added N often take climate as well as soil texture into consideration. The actual reduction in applied N based on amount of SOM should also consider the crop, since the time required to grow the crop and period of the year in which it is grown can have a bearing on the need for added N. In Arkansas, N recommendations for cotton vary with the location as well as with soil SOM. Recommendations call for an N reduction of about 10 lb/acre for each percent of SOM, but approximately 5 lb/acre more N are recommended for southern Arkansas than for northern Arkansas (Table 3.13).

Combining data on N release from SOM with information as to the previous crop and whether manure or other organic materials have been applied, along with the crop's ability to respond to N additions, promises to provide a better basis for recommending N applications.

TABLE 3.11. The effect of soil textural class on estimated N release from soil organic matter during the crop cycle

SOM (%)	N released (lb/acre)		
	Sandy loam	Silt loam	Clay loam
0.0-0.3	0-30	0-45	0-55
0.4-0.7	30-40	50-55	60-65
0.8-1.2	40-50	60-65	70-75
1.3-1.7	50-60	70-75	80-85
1.8-2.2	60-70	80-85	90-95
2.3-2.7	70-80	90-95	100-105
2.8-3.2	80-90	100-105	110-115
3.3-3.7	90-100	110-115	120-125
3.8-4.2	100-110	120-125	130-135
4.3-4.7	110-120	130-135	140-145
4.8-5.2	120-130	140-145	
5.3-5.7	130-140	150-155	
5.8-6.2	140-150	160-165	
6.3-6.7	150-160		
6.8-7.2	160-170		

Source: Ankerman, D. and R. Large. n.d. *Soil and Plant Analysis.* Memphis, TN: A & L Analytical Laboratories. Reproduced by permission of Dr. R. Large, A & L Analytical Laboratories, Inc., Memphis, TN.

A system using this approach for fertilizing 20 vegetables has been incorporated in the NPK Predictor, published by the Vegetable Research Trust in Warwick, England. To use the Predictor, they have established four indices, which reflect the different SOM and OM levels commonly found in different farming situations. A different level of N fertilization is suggested for each of the indices, and the level is modified according to crop requirements:

0 = Very low SOM soils previously farmed to at least two years of cereals

1 = Normal soil cropped one year with cereals or storage root crops

2 = Normal soil in arable or ley (temporarily under grass) systems

3 = Soils with large N reserves or that have received large applications of farmyard manure or other organic manures (composts, biosolids)

The amounts of N in kilograms per hectare (multiply by 0.8929 for lb/acre) to obtain top yields for a highly responsive crop, such as table beets, on the different indices are: 150 for #0, 120 for #1, 90 for #2,

TABLE 3.12. Recommendations for nitrogen topdressings of small grains grown in Indiana based on soil texture and yield goals

Soil texture	Yield goals (bu/acre)					
	Wheat or rye	30-44	45-54	55-64	65-74	75+
	Oats or barley	70-85	86-100	101-115	116-130	131+
		N (lb/acre)				
Sands		50	60	70	80	90
Sandy loams, loams		40	50	60	70	80
Silt loams, clays		30	40	50	60	70
Muck and organic soils		15	25	35	45	60

Source: Plant and Soil Test Laboratory, Agronomy Department, Purdue University, Lafayette, IN. Courtesy of Professor R. K. Stivers.

TABLE 3.13. Nitrogen recommendations for cotton as affected by soil organic matter

Soil OM (%)	Northern Arkansas (lb/acre)	Southern Arkansas (lb/acre)
3+	40-50	45-50
2-3	50-55	55-60
1-2	60-65	60-65
<1	65-70	70-80

Source: Miley, W. N. n.d. Fertilizing cotton with nitrogen. University of Arkansas Cooperative Extension Service Leaflet 526. Fayetteville, AK.

and 70 for #3. For a less responsive crop such as broad beans, 20 lb of N have been suggested for #0, 15 for #1, 10 for #2, and 10 for #3. (The Predictor can be obtained by contacting the Liaison Officer NVRS, Wellesbourne, Warwick, CV359EF, England.)

Basing N recommendations on SOM even when taking into consideration variations caused by climate, previous crop, soil class, and OM additions is subject to some problems. From the senior author's experience, estimated N release from SOM for recommending N fertilizer should be used with considerable caution. It can be helpful particularly if the soil is prepared for planting a few weeks beforehand, allowing the previous crop or other OM addition to undergo considerable decomposition. Its usefulness is greatly improved if it is combined with soil test data for available N (NH_4-N + NO_3-N) in soil samples collected shortly before planting and one or two weeks after planting. A later sample about midway in the season can also be helpful.

The senior author has had a number of experiences with SOM levels that theoretically should have supplied more N than was needed to grow an excellent crop, even if the lower estimates of release were used, but crop yields were seriously curtailed unless some readily available N was applied. These experiences evidently are not unique, since other workers have noted the difficulty of synchronizing nitrogen availability from organic sources with the demands of the crop (Pang and Letey 2000).

There are several possible reasons for these poor performances. A rather common reason is that conditions for nutrient release from the SOM are not favorable.

In the temperate zones, low soil temperatures are rather common for early planted crops. Because soil temperatures lag behind air temperatures, air temperatures may be satisfactory for crop growth, but cool soil temperatures inhibit N release (and often P release as well) from SOM, appreciably slowing crop growth. A substantial improvement in crop performance often can be made by a timely addition of N, rather than waiting for sufficient N to be released from the SOM. Adding P with the N may enhance this response.

Another cause for less than expected release of N is a lack of sufficient soil O_2, which is needed to decompose the SOM. Lack of soil O_2 is usually due to excess water or soil compaction. In humid zones, another reason for poor performance may be that heavy rains leach enough N so that fertilizer is needed soon after the rains to supply enough N for the crop to continue active growth until SOM decomposition can supply enough N.

To compensate for the problems of cool soils and N leaching, some N is recommended for various vegetable crops grown on organic soils. The high SOM levels in these soils should supply more than enough N without any fertilizer N, yet the University of Florida Cooperative Extension Service recommends using 30-40 lb nitrate-N/acre during cool periods or after a leaching rain for a number of vegetable crops grown on irrigated organic soils (Hochmuth and Hanlon 1995).

Another cause of poor performance based on estimated N release from SOM in soils with large amounts of SOM can be the simultaneous presence of large amounts of OM with wide C:N ratios or with sufficient polyphenols or lignins that slow the decomposition rate. Materials derived from crops, such as pastures or sugarcane grown for many years without major tillage, may consist largely of stems or roots with wide C:N ratios. Upon incorporating these materials, the decomposition of the large amounts of OM may temporarily rob most of the early released N from the SOM, even though the amounts released later from the SOM may be large and ample for later growth.

An example to illustrate this point probably occurred in Guatemala, as the company for which the senior author was acting as a consultant started farming in a new area previously devoted to sugarcane for a number of years. The SOM of the silt loam was about 4.6 percent, which theoretically should have supplied enough N to grow a good crop (600 boxes/acre) of cantaloupes. Using the lower estimate of N release from SOM of 20 lb/acre for each percent SOM should have supplied about 90 lb of N per acre, which is more than the amount of N being applied to soils with lower SOM values. On these soils with about 1.5 percent SOM, N applications are limited to about 60 lb/acre during the warm periods and 70 lb during the cool period of the year because larger amounts can yield excessively large sizes of cantaloupe that do not ship well. (In the warm periods, about 20 lb of N per acre are naturally supplied by the irrigation water but only about 10 lb/acre during cool periods.)

Although estimated N release was greater than the total N required for cantaloupes, we did not dare to eliminate all N, but tried reducing it by half. At the time, fertilizer was being applied through irrigation, and 25 of the total 60 lb N per acre for the crop was applied in the first week to provide for rapid early growth. Accordingly, the total N scheduled was reduced from the 60 lb/acre usually applied to 30 lb/acre for the new soils, with only about 12.5 lb/acre being applied in the first week.

Theoretically, the anticipated release from the SOM should have more than made up the difference needed to provide rapid growth, but early growth was much poorer on the high-SOM soil than the older soils with only about one-third of the SOM. The poor growth on the new soils with the higher level of SOM was readily explained by routine soil tests obtained about two weeks after planting, which showed an available N (NH_4-N + NO_3-N) content of 10-20 ppm in the high-SOM soils while the other soils had 30-50 ppm.

More N additions to the new soils improved early growth, but cantaloupe quality (size, netting, and sugar content) was universally poorer in fruit grown on the high-SOM soils. A possible explanation revealed by later soil tests is that much of the N released from SOM came rather late—four to eight weeks after planting. The late N addition spurred late vine growth at the expense of the fruit. After following soil test data from cantaloupe plantings for several years, it was concluded that the best yields with good quality are obtained if the available N in the first few weeks after planting is in the range of 50-75 ppm of soil. But this concentration needs to decline to a range of about 10-20 ppm at time of harvest lest quality be affected.

The late release of N from SOM in Guatemala was a surprise since temperatures during the plantings should have been ideal for early decomposition. Unlike early plantings in cooler climates, where soil temperatures are usually low and some N is needed at planting despite high levels of SOM, the failure of the SOM to provide sufficient N in Guatemala needs to be explained by other means. Although no definite answer is available, it is suspected that along with the high SOM, there were substantial amounts of OM (sugarcane debris and roots) with wide C:N ratios that seriously reduced the early release of N from the SOM.

The presence of large amounts of wide C:N ratio organic materials with high levels of SOM are not uncommon. The possibility that N is tied up as the wide C:N ratio materials decompose plus the poor release of N in cool soils make the use of estimated data for N release from SOM precarious unless there is a good evaluation of the available N by soil testing.

The problem of late N release from SOM makes soils having high levels of SOM unsuitable for growing fruit crops or other crops that are prized for their sugar or solids content. In addition to melons, it is expected that tomatoes, strawberries, sugar beets, cucumbers, and sweet potatoes will probably do poorly under such circumstances.

For crops that are not adversely affected by late release of N from SOM (such as snap beans, table beets, radish, lettuce, spinach, radicchio, kale, parsley, broccoli, cabbage, cauliflower, collards, endive, escarole, onions), N applications can be reduced and even eliminated on soils with high levels of SOM, providing that soil tests taken at planting and again at two weeks after planting show enough available N for the crop. If tests show inadequate N, just enough to provide an ideal N level for the crop must be added. The reduction of N applications on soils with high SOM levels is more apt to be successful if the preceding crop or OM addition was incorporated into the soil a few weeks before planting.

Although the poor correlation between SOM and crop yields calls into question the wisdom of using SOM contents as the sole interpreter for N recommendations, it should in no way denigrate the value of SOM as a source of nutrients for the primary crop. The nutrient contents of both OM and SOM can be used as a base for estimating fertilizer recommendations, but soil test results (primarily NO_3 and NH_4 concentrations) are required at critical times (cool soil temperatures, after leaching rains) and in the presence of compact soils to augment organic matter data. The need for the additional tests is greater with short-season annual crops, which need an almost uninterrupted nutrient supply for optimum yields, than it is for longer season crops (particularly perennials) that may have stored nutrient reserves that can tide them over lean periods. The need is also greater during certain times of the year (early spring in temperate zones, after rainy season in the tropics) and for short-season crops after incorporating large amounts of OM. The latter can occur as old pastures, hay, or sugarcane fields are converted to growing vegetable crops.

The poor association between quick test results for SOM and crop performance have prompted evaluation of other tests that may give a better evaluation of SOM as a provider of N. Nitrogen availability indices based on incubation of soil samples have been used. The long-term biological indices appear to be superior to the short-term tests in predicting N availability. Unfortunately, the senior author has found these incubation tests usually too time consuming to be of much practical value for a consultation service making fertilizer recommendations.

Because most of the quickly available N provided by the decomposition of OM comes from a small fraction of OM that is easily decomposed, it has been suggested that relatively simple procedures such as a mild acid or alkaline extraction could be used to determine these more readily available forms. Also, a procedure based on autoclaving a soil sample for 16 hours at 121°C (250°F) to release ammonia that can be readily measured has been offered as a means of determining the easily decomposed portion of organic matter that provides for early N release (Dahnke and Johnson 1990).

Although determining the easily decomposable portion of OM should help provide a better means of predicting N release than quick tests for SOM, it probably will not completely solve the problem of determining how much fertilizer N needs to be added to produce maximum economic yield (MEY) crops. Unfortunately, these methods cannot reveal how much N actually will be released, as the release still depends on temperature, moisture, and O_2 in the soil, all of which can vary a great deal in different years and with different soils and management practices.

Basing nitrogen applications on results of NO_3-N + NH_4-N in repeated samples has been used by the senior author for about 60 years to determine N needs of numerous annual plants with a great deal of success (Wolf 1982). Combining tests for the easily decomposable portion of organic matter with the tests for NO_3-N + NH_4-N may increase the accuracy of predicting N but the extra cost and time to run them may not be warranted.

Chapter 4

Physical Qualities

Both OM and SOM affect the physical nature of a soil. In so doing, nearly all factors that affect plant growth are modified. The erodibility of a soil is altered. The amounts of air, water, and nutrients available for plants and microorganisms change. The amount of water moving into a soil or the amounts of water and air stored or readily moving through it also change as various physical characteristics are modified by the organic matter. The amounts of nutrients released from soil minerals and organic materials and their availability for plants are also affected as OM changes a soil's physical nature, since it determines the ability of microorganisms and roots to function properly. Root growth, or the ability of a root to penetrate the soil, is also decided by size of soil pores, which are influenced by OM and SOM. By affecting soil aeration, physical qualities altered by organic matter affect the severity of several soil-borne diseases. Even the energy used to cultivate a soil is decided in large part by physical properties, which are improved by OM and SOM.

Physical characteristics of a soil may be modified by other means, such as tillage or applications of Ca, synthetic conditioners, or water, but none of these have the potential for beneficial effects on crop yields that are exhibited by the effects of OM. Although the nutrients supplied by organic matter can be obtained by other means, and usually at less cost, the physical benefits derived from OM are another matter. It is not possible, at least not economically possible at present, to duplicate all the physical benefits derived from OM by substituting other sources, making the judicious use of OM essential for the sustainability of agriculture.

Much of the effect of physical characteristics is designated as soil tilth, which, in essence, classifies a soil according to its ease of operation (seedbed preparation and planting) and chances for seedling and plant development, providing that water, nutrients, and climate are satisfactory. Other physical characteristics, such as those that affect air and water movement, water infiltration, storage, drainage, and ability of roots to penetrate soil, are also influenced by organic matter. Many of these attributes are largely dependent on soil structure.

The influence of organic matter on soil structure, an important soil physical characteristic, and several other soil physical properties that affect crop production are described under several headings.

SOIL STRUCTURE

Soil structure describes the manner in which the soil particles of sand, silt, and clay are arranged into larger units, and results from the tendency of fine particles of clay and humus to stick together. Various particles of sand, silt, and clay are cemented together to form peds. Although sand may remain as single soil particles, silt and clay particles tend to combine with humus and cementing substances to form larger units or peds, the grouping of which produces various soil structures. The larger peds are held together by cementing agents consisting of organic substances, iron oxides, carbonates, or silica. Several products formed from the decomposition of OM cement and stabilize the peds. Mucilages, probably polysaccharides, are absorbed on the surface, cementing the various particles together, while lipids derived from OM tend to waterproof them. The assemblage of the peds allows much better moisture holding capacity (MHC), water infiltration and drainage, and greater exchange of air and water, while lessening the dangers of compaction.

Soils may be structureless because (1) the units retain their individual characteristics (primarily sandy materials); (2) particles cling together as a single mass (typical of clay); or (3) structure has been destroyed because the soil was puddled or worked when too wet. But most soils do exhibit structure, the type of which affects the fertility of the soil by influencing the amounts of water, air, and nutrients that are available to plants.

Development of Structure

Structure develops through several processes, and organic matter plays a prominent role in most of these. At first, alternate freezing and thawing or alternate wetting and drying, with the resultant expansion or shrinkage, plus the destructive forces of wind and rain, break the rock mantle into smaller fragments. The mantle is further disintegrated by the action of microorganisms and plant roots. The latter tend to penetrate the various fragments, loosening the mass and leaving pore spaces as they decay. Organic matter greatly increases microbial action, which hastens the process and tends to promote conditions favorable for the introduction of insects, worms, and small animals. These organisms tend to hasten soil formation by moving organic matter deep into the soil, mixing the various fragments, and producing cementing agents that form peds. The mixing of mineral and organic matter is hastened by earthworms as they form casts, and by roots as they form channels capable of moving water and organic matter into the soil.

Types of Structure

In time, various forces result in different types of soil structure, depending on the nature of the parent materials, climatic factors, and type of organic materials. Essentially, the five different types of soil structure can be classified as follows:

1. *Platy structure:* Platelike aggregates, which are arranged around a horizontal plane, tend to crack horizontally. Puddling and ponding cause this type of structure at the surface, but lateral movement of water through the soil can cause it to form in any part of the soil. Continuous plowing at the same depth tends to develop the platelike plow sole or pan at the bottom of the furrow.
2. *Blocky structure* has units that are about as broad as they are long. Two types are formed. Angular blocky structure has peds with sharp corners and is usually found in the B horizon of humid region soils; subangular blocky peds have rounded corners and are found in either the A or B horizon. Blocky structure will exhibit both vertical and horizontal cracking. (Horizons are soil layers, approximately parallel to the soil surface. Elements are leached from the upper A horizon, many of them being deposited in the lower B horizon.)
3. *Prismatic structure,* usually found in subsoils, especially in dry regions, produces vertical rather than horizontal cracking and prismlike units that are much longer (two to five times) than broad.

4. *Columnar structure* also has units appreciably longer than broad, but the shape is approximately cylindrical as a column. It is usually found in subsoils with considerable sodium.
5. *Spheroidal structure,* characterized by units that have no sharp edges, is usually present in surface soils rich in OM and is considered the most suitable for producing crops. Two types of spheroidal structure are recognized. Granular structure consists of loosely packed granules that are largely separated from each other. Crumb structure, favored by farmers, is a form of spheroidal structure that is particularly porous. Both forms are aided by organic matter, but crumb structure is especially influenced by activities of microorganisms as they produce various gels and viscous products, capable of cementing the peds and stabilizing them. Fungi with their intertwining hyphae help to keep the crumbs open and porous.

The structure of the soil is greatly modified by (1) the amount and nature of OM and how it is handled; (2) soil tillage; and (3) additions of irrigation water, fertilizers, liming materials, gypsum, and synthetic long-chained organic materials. Organic matter usually is the single most important factor available to growers for producing and maintaining an ideal structure.

The Importance of Structure

Structure largely affects soil air and water (infiltration, drainage, and the amount of water held against gravitational forces) because of the spaces between aggregates, which are the pores or voids of the soil. The pores, largely dependent on the nature of the aggregates and how they are built into the structure, are also aided by channels or tunnels left by roots, rodents, insects, and earthworms and by the alternate shrinking and swelling of the clays.

Soil structure, both at the surface and internally, affects water infiltration. Surface crusts or capping can greatly restrict water entry. Compacted soil or hardpans slow water infiltration and drainage. If drainage is slowed, there will be insufficient exchange of air for excess water in time to avoid damage to the crop. More about infiltration and water movement in the soil is presented under the heading Infiltration and Percolation.

Many soil characteristics besides good aeration and water-holding capacity are influenced by soil structure. Good tilth, which allows easy seedbed preparation and planting, and increases the chances of rapid seedling and plant development, primarily depends on satisfactory soil structure. Good structure also provides rapid heat transfer, prevents moist soil from being excessively sticky, and allows it to retain much of its characteristics as it is worked while exerting a minimum of mechanical impedance.

The crumb or granular structure comes closest to fulfilling these requirements, especially if the crumbs in the surface soil are large enough that they do not blow away but are small enough to allow good contact with seed to ensure satisfactory germination. The effectiveness of this type of structure is largely due to the balance between macro- and micropores, which the crumb structure is able to maintain.

Factors Affecting Soil Structure

Structure in undisturbed soils is usually satisfactory. The results of weathering with alternate freezing and thawing or wetting and drying normally produce soils of good structure. Beneficial structure is enhanced by the presence of plants as roots penetrate soils, opening up channels, and by the decay of organic materials, which binds soil particles to provide better porosity. Poor structure in cultivated soils, however, is rather common, especially where high-tech monoculture is practiced. The single crop often limits the amount of organic materials returned to the soil, continuous cultivation usually associated with it helps deplete soil OM, and the heavy machinery usually employed in monoculture helps to unduly compact soils.

Nature is not always benign. While slow freezing of wet heavy soils benefits structure, rapid freezing, especially of sandy soils, may be detrimental. The improvement in structure that comes about from wetting and drying is related to the proportion and type of clay. Soils that contain large proportions of montmorillonite clay tend to change volume as the soil is wetted and dried due to swelling and shrinking. Cracks formed as the soil dries will remain for some time even though the soil swells as it is rewetted, and compaction tends to be a minor problem on such soils. The size of clods is reduced as the soil is gradually wetted, but rapid wetting or flooding can destroy structure.

Heavy rainfall or irrigation falling in large drops increase compaction and can cause capping of soils not protected by mulch. It is interesting to note that organic matter can be helpful in retaining suitable structure in the presence of rapid freezing of sandy soils, rapid wetting or flooding of clods, or the compaction from capping.

Structure can be affected by several other factors, such as the saturation of the CEC by various cations, how water is applied, and by chemical additions from irrigation water, fertilizers, liming materials, gypsum, and conditioners. Here too, organic matter may be helpful by reducing the force of applied water, increasing the CEC, and buffering and otherwise detoxifying the effects of adverse chemicals.

POROSITY

Porosity refers to that portion of a soil not occupied by solid particles and expressed as the ratio or percent of voids or pores to the total volume of the soil. In most fertile mineral soils, porosity is about 50 percent of the volume. Pores, or the voids of a soil, are formed as (1) irregularly shaped peds occupy a given volume of soil; (2) roots, insects, worms, and small animals push their way through the soil; and (3) several gases trapped in a water film expand.

Porosity is important because it influences

1. the amount of air present in a soil,
2. the rapidity with which spent air rich in CO_2 can be exchanged for air rich in O_2,
3. the amount of water held by a soil,
4. the rapidity with which water infiltrates a soil, and
5. how quickly excess water can be drained.

Good porosity helps not only with nutrient uptake by providing satisfactory air and water environments, but it also tends to increase the amount of nutrients available for uptake. The increase is made possible by the ideal air and water environments resulting from adequate porosity favorably affecting the number of microorganisms and their activities. These activities are responsible for the release of nutrients from OM, and the fixation of N both symbiotically or directly from the air.

The volume of total porosity in mineral soils ranges from 0.28 to 0.75 m^3m^{-3} and that of organic soils from 0.55 to 0.94 m^3m^{-3}. The optimum volume is probably about 0.50 m^3m^{-3}. Ideally, about half of this volume ought to be air porosity. As a minimum, about one-fourth of this porosity (0.12 m^3m^{-3}) ought to be present two to three days after irrigation or heavy rainfall.

A soil's pore space can be occupied by both gas and water. Different sized pores make up a soil's porosity, with most of the air being confined to the larger pore spaces, and is designated as air-filled porosity. The space occupied by water is primarily confined to the smaller pores and is referred to as water-filled or capillary porosity. The amount of readily exchangeable air is primarily dependent on the large pores, which regulate the rapidity with which spent air can be replaced with air rich in O_2.

Pore Size

The physical characteristics of a soil that permit rapid exchange of air and water and still retain enough water for good crop production depend on the soil having sufficient numbers of macropores (large pores) and micropores (small pores capable of capillary action). The macropores are useful for retaining and readily exchanging large amounts of air and also for the rapid infiltration of water and drainage of the excess. The micropores are useful for the retention and movement of water by capillary action.

Pores are classified according to their average diameter as being very fine (<0.5 mm), fine (0.5-2 mm), medium (2-5 mm), and coarse (>5 mm) (Miller and Donahue 1995). A slightly different classification lists macropores as being 0.08-5+ mm in diameter and micropores as having diameters of <0.0001-0.08 mm. The micropores are divided into different classes, with mesopores being 0.03-0.08 mm, micropores 0.005-0.03 mm, ultramicropores 0.0001-0.005 mm, and cryptopores being less than 0.0001 mm in diameter. All of the micropores retain water after drainage, but only the water in mesopore and micropore classes is available to plants. The water in ultramicropores and cryptopores is not available for plants, and the openings are so small that most microorganisms cannot enter (Brady and Weil 1999).

Coarse pores or macropores (>5 mm) extend around the main structural soil units. These pores, largely dependent on the nature of the soil particles or aggregates, are also aided by channels or tunnels left by roots, rodents, insects, and earthworms and by the alternate shrinking and swelling of the clays. In addition to sufficient numbers, macropores have to be present deep enough so that there is a rapid exchange of air in most of the root zone.

Sufficient numbers of macropores extending throughout the soil profile are also needed for water drainage. Rain or irrigation water needs to infiltrate the soil rapidly, but much of it also needs to drain rapidly so that the pores will regain sufficient air. Sufficient number and distribution of macropores ensures the removal of excess water and replacement with air. These pores are also instrumental in aiding the removal of air containing large amounts of CO_2 and little O_2 and replacing it with normal air containing much higher levels of O_2. Lack of O_2 in soils with insufficient large pore space can slow water and nutrient uptake and probably is becoming the limiting factor in many modern farm operations.

Retention and movement of water after initial drainage is made possible by micropores (small or fine pores <0.08 mm), which extend through the interior of the soil structural units. There must be sufficient numbers of them to hold enough water against drainage forces and so located that they are readily reached by plant roots. The small pores or capillaries not only hold water but can move it upward from lower layers. A sufficient number of them, properly handled, allow dryland farming. Movement of water in the micropores is practically limited to capillary action. A soil with insufficient micropores tends to be droughty and unproductive unless there are ample rains, although the recent use of irrigation, especially drip irrigation, has enabled efficient use of such soils.

Desirable Porosity

To maintain good growing conditions for most crops, it is necessary to have soil porosity, which will allow for

1. rapid movement of air and water into a soil,
2. rapid exchange of air and water,
3. rapid exchange of spent soil air for fresh air,

4. ready removal of excess water,
5. sufficiently large pores for root entry, and
6. sufficient number of small pores to hold water against gravity but not so small that the water is not available for plants.

Air and Water Entry

The entry of soil air as well as water is primarily dependent on the presence of enough large pores near the surface and extending deep into the profile. In some soils, cracks or channels left from roots or burrowing animals can aid the process. The absence of crusting or compact surface aids both air and water entry. Organic matter, particularly when it is left as mulch, tends to provide and protect favorable conditions for air and water entry.

Air and Water Exchange

Since either water or air can occupy the same pore space, all pore spaces may be filled with air and/or water under field conditions. If most large pores are entirely filled with water, as can happen after a rain or irrigation, O_2 will be insufficient for needed respiration. It is important that there are sufficient interconnected large pores so that water can be readily removed, allowing air again to occupy about half of the total pore spaces.

Even though the soil or medium may have ample porosity, problems arise at times with interconnecting pathways for rapid exchange of air and water, primarily because a large proportion of the pores is filled with water from too much rain or irrigation. The problem may exist for a short time and usually will have no perceptible effect on crop yield if sufficient large pore spaces are present and the excess water can readily drain out of the root zone. If, on the other hand, large pore space is in short supply or if poor drainage conditions due to compact subsoils, plow soles, or hardened layers occur, or if there is a high water table, water movement out of the root zone may be much too slow. Slow water movement can seriously impinge on the replacement of CO_2 and other gases harmful to respiration with air rich in O_2, needed for respiration.

Rapid Exchange of Spent Air for Air Rich in Oxygen

Sufficient porosity for a rapid exchange of soil air is essential for normal growth of roots and microorganisms. Carbon dioxide from the respiration of microorganisms and plants tends to increase in soil air. Sometimes other soil gases that are produced under anaerobic conditions, such as methane, ethylene, nitric oxide, and hydrogen sulfide, will be present in addition to the CO_2. This air needs to be exchanged for air rich in O_2, lest the respiration of microorganisms and plant roots be seriously hindered.

Most of the exchange of soil air takes place in the air-filled pore space by the process of diffusion, although mass flow resulting from temperature or pressure gradients that increase the volume of air-filled pore space as water is withdrawn may be responsible for some of the exchange.

Although much smaller in amount, there is an important exchange of O_2 and CO_2 in water-filled pores. The small amount of O_2 in water-filled pores is due to the poor solubility of O_2 in water (0.6 percent) as compared to the normal presence of 21 percent O_2 in air. Also, the diffusion of O_2 in water-filled pores is only 1/10,000 that in air, making its movement rather limited. Despite the small amount of O_2 in water-filled pores, its presence ensures sufficient O_2 for microorganisms and root hairs since these need to be bathed in a water film for maximum development. Failure to maintain sufficient O_2 in these films leads to decreased growth and possible death of cells.

Just how much air needs to be exchanged to maintain good O_2 balance in the soil or medium varies with the activities of plants and microorganisms, which are affected by (1) temperature and moisture; (2) the kind of plant, its age and vigor; and (3) microbial populations. On average, it has been estimated that the O_2 consumption is about 10 liters per m^2 of soil per day. The 10 liters of O_2 are equivalent to about 50 liters of air or slightly more than 53,000 gallons of air per acre—a tremendous amount of air that needs to be exchanged in order to maintain healthy conditions for optimum crop production. If the air porosity of the soil is 20 percent, there will be 200 liters of air in a cubic meter of soil. If the 50 liter consumption rate is uniform to a depth of one meter, all the O_2 would be consumed in four days, requiring complete replacement of the 200,000+ liters (53,000 gallons) in that time. In most fertile soils with adequate pore space, the exchange of O_2 for built-up levels of CO_2 is readily accomplished. But the consequences are far different and results in greatly reduced crop production if air-filled porosity is compromised because of inadequate pore space due to compaction or deflocculation or the presence of water that does not readily drain.

The critical content of porosity in open soils begins at about 5-10 percent by volume, with root growth of many plants being limited when O_2 is less than 0.1 percent of the pore space (Glinski and Lipiec 1990). The optimum volume in open soils is probably about 0.25 m^3m^{-3}. As a bare minimum, about half of this porosity (0.12 m^3m^{-3}) ought to be present two to three days after irrigation or heavy rainfall. It has been noted that the health of conifers is greatly reduced as soil air content falls below 4 percent, that of deciduous trees and fruit trees if it falls below 10 percent, and that of nurseries and some vegetables if the air content falls below 15 percent of the total porosity for relatively short periods.

Air-filled porosity is of more critical concern in pot cultures, because of the small volume of soil and a perched water table. While some plants and short-term growing (seed germination) can be productive in media with rather low air porosity (2-5 percent of volume), some container-grown plants (azaleas and orchids) require media with very high air porosities (20 percent of volume), and many other plants require porosities of 10-20 percent.

Porosity and Root Entry

Roots tend to penetrate soils by growing through existing pore spaces or cracks in the soil or channels left from preexisting roots. Root growth is inhibited if the pore spaces are too small for root entry unless there are cracks or channels in the soil. The size of the root tip varies with different plants, making it easier for some plants to penetrate soils with limited pore space. If the open space is not large enough, the root tips will tend to elongate by moving aside soil particles. Elongation of the root tip is adversely affected by soil strength, which increases as compaction or bulk density increases.

Porosity of Different Soils

The amount of air-filled porosity is not closely related to total porosity because of the variable water content. Total porosity is usually greater in finely textured soil, with the clays having the most and coarse sands the least. Although clay loams and clays have a high total air space, air porosity suitable for rapid air exchange may be low. Sands, which have less porosity, usually are better aerated than the finer-textured soils. Porosity in sands consists primarily of large pores, while the clays have small pores unless a large percentage of the particles have been built up into peds. At times, much of the pore space in clays is filled with water because pores or interconnecting passages are too small to allow rapid and efficient drainage. On the other hand, despite lower total porosity, air-filled porosity and air exchange are relatively high in most sands, where large particles allow large air spaces that are readily drained and refilled with air. For effective air exchange, it must be remembered that the large pore spaces need to be adequately connected by passageways at least 0.01 mm in diameter.

Exceptions to the high air porosity of sands occur in very fine sands or coarser sands that may have restrictive subsoils limiting water drainage. Air spaces between very fine sand particles also may be small enough to limit water drainage.

The amount of air-filled porosity in the finer-textured soils can be quite variable, because clay and silt particles tend to combine with other particles to form various aggregates. The amount of air-filled porosity is largely dependent on the nature of the aggregates and how well they can be maintained. Unless a fine-textured soil has a large number of stable aggregates or channels and cracks, it will tend to have insufficient air-filled porosity.

Maintaining Satisfactory Porosity

As will be pointed out in later chapters, one of the more important practices for improving porosity is the judicious use of sufficient organic matter, although porosity can be improved, at least temporarily, by proper cultivation, the addition of calcium, or the use of synthetic large-molecule polyelectrolytes.

Generally, cultivation can benefit air porosity temporarily, but it usually leads to the formation of smaller pores at the expense of large pores with a negative effect on air exchange, water infiltration, and drainage. Tillage of soils when wet is highly counterproductive, but tillage of slightly frozen soils can aid in their drying, allowing earlier spring planting. Successful soil drying by frozen tillage requires that the frozen layer be 1-4 inches deep without snow accumulation (Gooch 1997).

The addition of calcium in the form of limestones or gypsum, or the addition of large-molecule polyelectrolytes can be helpful on finer-textured soils. Based on the senior author's experience, the coarse-textured soils (sands) fail to show substantial porosity changes from limestone, gypsum, or polyelectrolyte additions.

BULK DENSITY

One of the benefits of good soil structure is that it lowers bulk density, which is the weight of soil solids per unit volume of soil and is obtained by dividing dry soil weight by its original volume. Variation in bulk densities is the result of differences in soil texture, organic matter contents, and management practices. Lower bulk density is associated with better aeration but also with a lower weight of soils. The latter property is important as soils are moved or worked. The reduced amount of power used to work soils with low density can lessen damage to soil structure.

The common bulk densities of mineral soils range between 1.0-1.6 g/cm^3, with the finer-textured soils having lower bulk densities because of their greater pore spaces. If there were no pore spaces, the bulk density of soil would be 2.65 g/cm^3 (the average density of soil mineral matter). Soils with low densities (1.0-1.4 g/cm^3) offer little impediment to the extension of roots and allow rapid movement of air and water. The bulk densities of organic soils are below 1.0 and those of the sphagnum moss peats will be close to 0.1 g/cm^3. Organic matter lowers bulk density because it is much lighter in weight than a similar volume of mineral matter. Organic matter also increases the aggregate stability of a soil, which tends to benefit pore space, lowering bulk density. Common bulk density values for several soil textural classes are given in Table 4.1.

Bulk density increases as soil is compacted but is improved by practices that increase aeration. Tillage usually lowers bulk density—

TABLE 4.1. Average bulk densities of different soil textural classes

Soil texture	Bulk density (g/cm^3)
Sand	1.58
Loamy sand	1.52
Sandy loam	1.47
Sandy clay loam	1.44
Silty clay loam	1.40
Loam	1.39
Silt loam	1.36
Sandy clay	1.33
Clay loam	1.31
Silty clay	1.26
Clay	1.23

Source: Information Capsule #137, Midwest Laboratories, Inc., Omaha, NE. Reproduced by permission of Dr. Kenneth Pohlman, Midwest Laboratories, Inc., Omaha, NE.

at least in the short term. Subsoiling lowers bulk density, but puddling the soil tends to increase it.

Generally, crop yields tend to improve as bulk densities are lowered, probably because high bulk density restricts root growth. Root growth tends to be slowed in soils with bulk densities 1.5-1.6 g/cm^3 and usually stops as bulk densities reach the 1.7-1.9 g/cm^3 range (Thompson and Troeh 1978). Usually root length decreases as bulk density increases, but absolute values needed for optimum growth are difficult to define because the optimum values may vary with different crops and different soil textures. Tolerance to soil compaction, which tends to increase bulk density, is in the order alfalfa → corn → soybeans → sugar beets → dry edible beans (Miller and Donahue 1995). While a bulk density value of <1.4 g/cm^3 is satisfactory for many crops on clay soils, a slightly higher value of <1.6 g/cm^3 can be tolerated on sands.

Bulk density has an effect on the movement of nutrients by altering their diffusion rate. The bulk densities at which diffusion is optimum probably vary with different soils, but the upper limits appear to be in the range of about 1.3-1.5 g/cm^3. Normally, soil moisture moves in a convoluted path around solid particles in the soil. As a very open soil is slightly compacted, the volumetric soil moisture content increases and diffusion is increased by the continuity of the water in the capillaries and as the length of the diffusion path is shortened. But as the soil is further compacted, the solid particles come together, making the diffusion path more tortuous as moisture must move around the particles.

In addition to the use of extra organic matter, the benefits of lower bulk density can be achieved by the use of subsoilers and some other forms of tillage. The tillage effects are useful for relatively short periods and do not come close to providing all the benefits of organic matter.

Relation of Bulk Density to Porosity

There is a close interrelationship between bulk density and porosity, with porosity decreasing as bulk density increases. The lower porosity provides poor aeration, which often is associated with reduced plant growth and, at times, may be related to certain soilborne plant diseases. Soil density values greater than 1.2 g/cm^3 on loams of Washington State were associated with higher incidence of bean and pea diseases (Miller and Donahue 1995).

COMPACTION

Compacting a soil can increase its bulk density and reduce the numbers of large pores, which in most cases results in reduced root activity. Some compaction of excessively open soils can be beneficial by improving capillary movement and diffusion of water, and making better contact of seed and soil. But compaction beyond a certain point can reduce a soil's ability to absorb water and provide sufficient O_2 to maintain healthy roots. The reduced absorption increases the possibility of serious erosion. Lack of O_2 reduces respiration of roots and microorganisms and increases the potential damage of roots from several soilborne diseases. Compacted soil is more resistant to root penetration, usually limiting the extent of the root system and making the plant more subject to shortages of moisture and nutrients. Compacted soils also require more energy for tillage operations.

Root length tends to be inversely related to compaction, and such roots are usually irregularly shaped. Much of this change in root growth appears to be due to differences in the oxygen diffusion rate (ODR), since compaction alters the ODR. The ODR in compacted soil can fall to less than 33 mg·m^{-2}·s^{-1}, and root elongation is seriously limited as the ODR falls below 58 mg·m^{-2}·s^{-1} (Erickson and Van Doren 1960).

Compaction can be present in various portions of the soil. Surface crusting and topsoil compaction are often present, but hardpans made up of densely packed sediments can exist in subsurface layers. Some can be cemented to the point of being rocklike (fragipans or ortstein). Claypans, having a very high clay content, that are resistant to penetration of air and water when wet may be present as subsoil layers.

Compaction can result from both natural and artificial forces, but soils may be affected differently by these forces. Soils with appreciable clay and little organic matter are markedly affected. Fine-textured soils with little montmorillonite clay are especially subject to compaction.

Although some compaction occurs naturally (rainfall impact and subsoil compaction due to load of topsoil), much of it is a result of modern farming operations. Topsoil compaction results from compression of foot traffic, extensive animal grazing, or use of machinery. The extent of crusting increases as the soil is deprived of cover or if it contains little OM to stabilize the soil aggregates. Many farm practices involving machinery, such as soil preparation, spreading of fertilizer and lime, cultivation, spraying, and harvesting, can seriously compact soils. Vehicular movement, particularly on wet, finely textured soils, compresses topsoil, although some of the force may persist to the subsoil. The effect of these forces plus slicks produced by plows and disks often produce compact layers or soles that can seriously affect water drainage and restrict root penetration.

Various mechanical methods can be used to limit compaction, such as (1) reducing weight of machinery; (2) limiting use of machinery when soils are wet; (3) confining spraying and harvesting machinery to special roadways (use of long booms limits amount of ground lost to production); and (4) spreading the weight of machinery over larger areas by use of floater tires, crawler tracks, and all-wheel drive. Damage resulting from overhead irrigation can be reduced by using nozzles that provide small droplets. Improving internal compaction problems by use of subsoilers is usually of short-term value, unless the subsoiling is combined with the addition of gypsum and/or organic matter.

While every effort should be made to reduce compaction from use of machinery or irrigation, it must be remembered that proper use of organic matter can limit much of the damage. The presence of sods or mulches substantially reduces the effects of rainfall or irrigation. Surface compaction can also be alleviated to a large extent by increasing stability of surface aggregates with such farming practices as suitable rotations and use of limestone or gypsum. Using rotations that increase OM also reduces the harmful effects of heavy machinery, and leaving organic residues on the surface to break the force of rain or irrigation droplets can reduce surface compaction. (The importance of OM from rotations and its special place as mulch to reduce compaction are covered more extensively in Chapter 6.)

INFILTRATION AND PERCOLATION

Infiltration, or the entry of water into a soil, and percolation, or the downward movement of water in a soil, are markedly influenced by OM and SOM.

The infiltration rate or capacity is dependent upon soil slope, presence of mulch, rate of water application, soil texture, soil structure, sequence of soil layers, depth of water table, moisture content, chemical content of the water, and length of time that water has been applied. The infiltration rate is much higher when rain starts to fall or irrigation is first applied but decreases with time. At first, the rate is influenced by soil properties and the tension existing between a dry or nearly dry surface and the wetter layers below. The difference in tension between the surface and lower layers becomes miniscule as the soil is wetted, and infiltration rate is largely influenced by gravity, although water conductivity of the soil can affect infiltration rate. The infiltration rate is usually lessened with time as the pore sizes are altered with partial deflocculation and breakdown of the peds by water action, but eventually the rate reaches an equilibrium that remains rather constant.

Initial infiltration is greater if the soil is not wet, but initial and later infiltration are superior if there is no surface crusting and there is an abundance of large pores near the surface. In the tropics and semi-tropics, the aggregates are of mineral origin as hydrous oxides combined with kaolinitic clay. These tend to be stable and probably account for the excellent infiltration on many tropical soils. In temperate zones, the stability of the aggregates is largely decided by organic matter, which can be readily depleted under intensive cultivation, leading to poor infiltration on many soils.

Generally, infiltration rates decrease as soil moisture rises, as happens with continued rainfall or irrigation, but some soils can become so dry that it is difficult to wet them. Some of the slowdown due to continued watering is due to collapse of some large pores as the soil is wetted. Large pores formed by tillage are especially prone to this type of collapse. Raindrops or drops of irrigation, as they impact on the soil surface, also tend to slow infiltration. The force of the impact tends to detach soil particles, producing a crust. Also, the detached particles move downward, plugging pores, and so reduce the rate. The harmful effects of large drops from overhead irrigation can be lessened by using different types of irrigation, or changing nozzles to supply smaller drops. Mulches, or large amounts of crop residues on the surface, also can be helpful in maintaining good infiltration rates by breaking up the droplet size.

Infiltration can be impeded at times by certain surfactants or repellents, which may be natural in some soils or may result when organic substances derived from citrus or grasses break down. Some repellency appears to occur after burning of brush. The problem can be serious as it limits the amount of water available for plants, and it can increase erosion. Control is possible because in most cases, the material is close to the surface, and cultivation usually can open enough surface for sufficient infiltration.

Infiltration of water can vary from less than 0.1 in/hour to more than 5 in/hour. Common infiltration rates are in the range of 0.3-1.0 in/hour. Values <0.2 in/hour are extremely slow. Such slow infiltration can lead to wasted water as runoff or ponded water, which can cause serious problems of soil crusting and poor aeration. Rates of 2-5 in/hour are very rapid, signifying poor water retention and waste of water by excessive percolation.

Best water infiltration occurs on level or near level land. If land is sloped, contour farming aids infiltration. Infiltration of water in dispersed soils or those with considerable slope or poor infiltration due to surface crusting is still practical if water is applied very slowly, as is possible with drip irrigation.

Very low or high salt content in irrigation water impedes infiltration into soils with appreciable silt and clay. Water with very low calcium, high sodium, or high total salts tends to move very slowly into heavy soils. Remedies include adding calcium salts to irrigation waters low in calcium or total salts, and gypsum to soils for better penetration of water high in salts.

Percolation is primarily dependent on continuous pore spaces in the soil. The flow rate is greatly affected by the size of the continuous pores, increasing about four times for every doubling of the pore diameter. Because of this dependence, it is closely related to soil texture, increasing with coarse texture, good structure, increased amounts of organic matter, and absence of compaction. Other factors that influence its rate are increased depth of soil, decreased soil moisture (up to a point), and increased soil temperature. Generally, sandy soils will have better percolation rates than heavier soils, but fine-textured soils that have ample organic matter, good structure, and well-developed stable aggregates will also allow rapid water movement—often more rapidly than fine sands.

Water permeability slows whenever there are contrasting layers, such as a clay layer below a sandier, more open layer, as is often found in the B horizon. Plow soles or other dense, compact layers or even bedrock may also slow infiltration, increasing the likelihood of overly wet areas, even to the point of ponding.

Slow permeability can be improved with better soil structure, using low-salt water, adding composts or manure, and removing excess salts. Excessive permeability is remedied by the addition of manure or other forms of organic matter or by barriers placed at strategic depths.

A more serious problem exists if percolation is restricted due to insufficient porosity, the presence of impermeable layers, such as fragipans, hardpans, plowpans, clay horizons, and clayplans, or the presence of a high water table. Some of these conditions can be modified by increasing organic matter, reducing compaction, deep plowing or using subsoiling to break up hard layers, and installing proper drainage.

Introducing additional organic matter often can improve poor percolation that is not due to impermeable layers or high water tables. Poor percolation was one of the first indications of poor growing conditions in Guatemala due to reduction in SOM (experienced by the senior author and commented upon earlier). Water, applied by drip irrigation, tended to remain in a small volume of soil immediately below and at both sides of the drips, with very little horizontal movement. Soil directly beneath the drip openings was saturated while the sides of the bed, 2.5 ft from the drips, were totally dry. Water movement, laterally and vertically, has greatly improved since a sorghum cover crop has been included between melon crops. Improvement was noticeable in the first melon crop following the cover crop and further improvements have been noted as three cover crops have been used. The change in permeability has made it easier to manage the water application and has resulted in better melon yields and quality.

AGGREGATE STABILITY

As pointed out, many soil physical characteristics are improved by the presence of aggregates. Porosity particularly is improved by the presence of large numbers of aggregates providing considerable large pores. How well porosity and other physical properties such as infiltration and permeability are maintained depends not only on the number and size of these aggregates but their stability, or how well they resist disintegration by water.

Wet sieving is often used to determine soil aggregate stability. A sample of soil is placed on top of a series of sieves with openings of decreasing size and capable of measuring aggregates with diameters 5-2 mm, 2-1 mm, 1-0.2 mm, and less than 0.2 mm. The different sizes are separated by slowly raising and lowering the sieves in water for a given length of time and measuring by weighing the dried amounts retained on the different sieves. The results are usually expressed as percentages of the total aggregates that are larger than 2 mm. These larger sizes are considered to be of great importance in the expression of the various physical qualities.

Stability of the aggregates depends on the cementing agent. In most soils, organic substances, along with silica, iron oxides, and carbonates, are the principle cementing agents, but phosphoric acid, sodium and potassium silicates, and synthetic long-chain polymers have been known to impart stability to soil aggregates. Studies in the Netherlands have shown that the stability of soil aggregates is higher in soils with higher organic matter contents and with greater numbers of earthworms. The number of earthworms was very important in maintaining the stability of macroaggregates. Stability was negatively correlated with conventional agriculture, evidently because of its use of synthetic fertilizers and pesticides, which may have limited earthworm activity (Brussaard 1997).

Aggregates evidently form through the addition of polysaccharides derived from organic matter, but their stability may depend on the intertwinement of hyphae in the aggregate. Fungi, concentrated near the surface in mineral soils under no-till, are very helpful in stabilizing aggregates formed by earthworms and microarthropods, such as mites (Beare 1997).

Although earthworm activity is very important in maintaining aggregate stability, fungi and bacteria also contribute directly to the formation and stability of soil aggregates. The contribution of fungi and bacteria often becomes dominant in conventional tillage as the activities of earthworms and other arthropods are often reduced because of tillage, lessened SOM, and use of certain fertilizers and insecticides.

Studies indicate that different organic materials may have different effects in stabilizing the aggregates (Piccolo et al. 1997). These studies point out that the composition of the organic matter, especially the humified portions, may have an important bearing on the efficiency of organic matter in stabilizing the aggregates. More recent work (Piccolo and Mbagwu 1999) suggests that the addition of organic materials rich in hydrophobic components (substances that resist water) will have a greater beneficial effect on the stability of aggregates than the addition of organic materials with predominantly hydrophilic components (substances that absorb water), such as polysaccharides. According to this and the previously cited work of Piccolo and colleagues, humic acid, because of its rich content of hydrophobic components, could be used to stabilize aggregates. Humic acid can be readily derived from relatively abundant sources, such as lignite or oxidized coal and, therefore, probably could be used economically to revitalize soils with poor physical qualities.

MOISTURE-HOLDING CAPACITY

The amount of water that a soil can hold against the downward pull of gravity is known as its moisture-holding capacity (MHC). This amount is affected by the presence of clay and organic matter. There is a close relationship between soil textural classes and MHC (Table 4.2).

Although all of the moisture is not available to plants, knowledge of the maximum amount held by the soil is helpful because it is indicative of the potentially available water and how much water can be added before a soil is saturated.

Only a portion of the total MHC is available for absorption by plants. The maximum amount of water that a soil can hold against gravity after it has been saturated and free drainage has ceased is called field capacity. The portion of water held between field capacity and the wilting point (when plants wilt permanently) is available for plants. All hygroscopic water and part of the capillary water held by the soil at tensions greater than 15 atmospheres is not available to most plants.

Moisture held between field capacity and permanent wilting also increases with increased amounts of SOM and, to a large extent, the clay content of the soil. Usually, available moisture increases with increasing clay content of the textural class, but amounts held at the permanent wilting points of some clays are so large that less moisture is available for plants (Table 4.3).

Note that clay loam and clay soils have appreciably less available water than loam. Although clay soil has very high field capacity, its smaller amount of available water in comparison to loam is due in part to a much larger amount of water held at the wilting point.

TABLE 4.2. The relationship between soil textural classes and moisture-holding capacity

Textural class	MHC (% volume)
Sand	38
Loamy sand	40
Sandy loam	42
Silt loam	45
Clay loam	50
Clay	60

Source: Modification of a table by Hillel, D. 1987. *The Efficient Use of Water in Irrigation.* World Bank Technical Paper #364. Washington, DC: The World Bank.

TABLE 4.3. Available moisture of different soil textural classes as affected by field capacity and permanent wilting points

	Soil moisture content		
Textural class	Field capacity (%)	Wilting point (%)	Available water (%)
Sand	9	2	7
Loamy sand	14	4	10
Sandy loam	23	9	14
Loam	34	12	22
Clay loam	30	16	14
Clay	38	24	14

Source: Boswell, M. J. 1990. *Micro-irrigation Manual.* El Cajon, CA: Hardie Irrigation.

Not all soil water held between field capacity and the wilting point is equally available to plants, because of its position or because it is held by attraction to other molecules of water and to the soil, the latter of which increases as the amounts of water decrease.

The failure of plants to take up water due to the position of the water can be modified by large, effective root systems and the presence of sufficient micropores that help move water to the roots by capillary action.

The force by which water is held in soils increases as the amount of water decreases, making it more difficult for plants to take up water as amounts decrease. This force has been quantified as atmospheres of pressure or atmospheres of suction (tension) or negative pressure necessary to remove the water. Water in a saturated soil is held at zero atmospheres but increases to about one-third atmosphere at field capacity, to 15 atmospheres at permanent wilting, about 1,000 atmospheres in an air-dry soil, and 10,000 atmospheres in oven-dry soil. These would be zero bars suction at saturation, 0.338 bars at field capacity, 15.1 at the wilting point, 1,013.3 for air-dried soil, and 10,133 for oven-dried soils. It is obvious that soils with large moisture-holding capacities that are largely filled with water have an initial advantage in supplying large quantities of water before the forces holding the water become so great that plants have great difficulty in removing it.

Organic matter can absorb considerable moisture, making it a logical approach for improving MHC. Various plant parts can absorb two to three times their weight of water. Humus can absorb 4-6 parts of water and peats will hold 9-25 parts of water, whereas clay is saturated with about 0.5 part of water by weight (Waksman 1938).

The need for increasing MHC is greatest in sandy soils, which have much smaller MHC than soils with finer textures. An increase of about 1 percent SOM can add about 1.5 percent additional moisture by volume at field capacity, but, as has been pointed out earlier, increasing SOM is a formidable task unless there are substantial changes in cultural practices along with the OM additions.

EROSION

Several of the qualities of OM that enhance the physical properties of a soil, such as infiltration and percolation, structure, aggregate stability, and compaction, reduce the erodibility of soils. The influence of OM in reducing erosion could be one of the most important characteristics of SOM in maintaining crop production and thereby sustaining agriculture.

The importance of reducing erosion is evident when one realizes that in 1986, it was estimated that about 2.5 billion tons of topsoil were being lost from the nation's croplands each year by both wind and water erosion. This loss is equivalent to about 5.0 tons/acre, with about two-thirds of this being lost to water and one-third to wind. It has been estimated that some cultivated soils are eroded at ten times the average rate (Brady and Weil 1999). These losses, particularly those exceeding the average, pose serious problems for agriculture, indicating that too many soils are losing productivity at an alarming rate.

Water Erosion

Erosion by water is affected by the extent of soil covered by OM, since OM protects the soil from splash erosion, whereby the soil aggregates tend to break down from the impact of raindrops or overhead irrigation drops. The presence of OM tends to reduce the destructive nature of these drops, as it absorbs much of their kinetic energy. The breakdown of the aggregates tends to form a crust, further reducing the ability of the soil to absorb the water. The soil dispersed by the action of rain or overhead irrigation drops is swept away by the increased runoff, which at first moves as a sheet, but soon forms small channels (rills), which develop into larger gullies that are capable of moving more soil and water ever faster. Although the formation of gullies is the most dramatic, often making such areas unfit for cultivation, rill erosion can also dramatically reduce productivity of soil because of the rapid removal of topsoil. Sheet erosion, although less noticeable than either rill or gully erosion, also cuts into production because it tends to remove the upper part of the soil, which is the richest, as it has the most biological diversity and contains the greatest amount of nutrients and SOM.

The extent of erosion loss depends upon several factors, such as the amount of rain and its intensity, the nature of the soil, grade and extent of slope, soil management including degree of tillage, amount of surface residue, canopy cover, terraces or other devices to slow water movement, and whether crops are planted on the contour.

The amount of rain or irrigation, while important, is much less so than the rate at which it is delivered. A 6-inch rain extending over a couple of days may cause little erosion, but the same amount occurring in an hour or two, as often happens in the tropics, can be devastating. The amount of erosion accelerates rapidly as the slope increases. Steep slopes tend to increase erosion because they increase the rate of water movement, thereby increasing its destructive power.

Soil texture, OM content, and presence of hardpans or compact layers can influence the rate of water infiltration, and thereby affect ponding or surface accumulation of water. The more open soils (coarse sands or finer-textured soils with ample SOM and free of hardpans or compact layers) have ample porosity for rapid infiltration, thereby lessening the amount accumulating on the surface. Water accumulating on the surface accelerates erosion by disintegrating aggregates and loosening soil, which readily moves at increased speed as the extra volume intensifies its destructive nature.

Management practices that limit tillage and increase the amount of ground cover tend to reduce erosion. Practices that promote rapid growth to supply early canopy cover can be helpful. Monoculture, or use of short rotations, tends to intensify losses from erosion. The longer rotations, with their tendency to include legumes, promote reduced tillage, thereby increasing organic matter and reducing erosion. A case is cited with extremely different water erosion losses of 13.1

tons of soil per acre per year from continuous winter wheat as compared to only 2.4 tons of soil per acre per year from areas in winter wheat but using leguminous meadows in the rotation (Reganold et al. 1987).

Soils that are recently cultivated and without organic residues on the surface, or that lack canopy cover, can be severely damaged by heavy rainfall and furrow or overhead irrigation. However, even on these soils, any practice that slows water movement can be beneficial. Water movement can be slowed by (1) planting on the contour, as is done in strip cropping; (2) providing contoured grass waterways to move water safely; and (3) building terraces to slow water movement by collecting it behind the terrace and channeling it to grassed waterways that can slowly move the water downhill to a body of water.

Erosion losses affect more than the fields being eroded. Soil and various kind of debris move downhill, often impinging adversely on fields below, at times dislodging seeds and plants or inundating crops with enough silt and sand so that they may not be able to function normally. Losses of productivity may continue for many years, because normal air and water movement in the covered soil can be altered beyond quick repair. The damage extends not only to soils below the eroded soil but also to various bodies of water, as silt and debris damage the quality of pond irrigation water, clog waterways and drainage ditches, and often seriously affect marine life by altering effective light penetration and oxygen levels in the water.

Wind Erosion

Wind erosion can also cause serious losses of soil. A considerable amount of wind erosion is due to direct wind pressure in regions that have few or no windbreaks. Most of this type of damage occurs on sandy and muck soils. Besides loss of soil, such movement can cause serious damage to young seedlings. In addition to direct movement of particles about 0.1-0.5 mm in diameter, wind can move particles greater than 0.5 mm in diameter by rolling them in a surface creep. Particles smaller than 0.1 mm in diameter are readily moved by wind, as they can be suspended in the air and moved great distances.

Although usually accounting for a small weight of the soil moved by wind, the portion suspended in the air may be most harmful in the long run, because this portion is rich in fine organic matter particles, the loss of which may have the most effect on a soil's properties.

Chapter 5

Biological Effects of Organic Matter

The importance of organic matter, as both OM and SOM, in supporting a biological population in the soil is immense. It is the biological influences of organic matter that distinguish a soil from a mass of inert rock fragments. As with physical effects derived from organic matter, it is difficult, if not impossible, to duplicate all of the biological effects of organic matter by other means.

We have touched upon some of these effects in describing the release of nutrients as OM decomposes, the fixation of N by both symbiotic and nonsymbiotic organisms, and the enormous changes in the physical quality of soils caused by organic matter. In this chapter, we expand on these items as well as cover in some detail several other important influences of organic matter upon biological properties that affect the health of soil and plants. Some of these effects, such as suppression of plant diseases, are beneficial, but others can depress crop yields by increasing the number of weeds or by harboring certain insects, disease organisms, or nematodes.

SOIL ORGANISMS AND THEIR FUNCTIONS

Organic matter has a profound effect on the number and kinds of organisms that are present in a soil. These organisms, consisting of microflora and microfauna, do not just live passively in a soil performing their own functions but are affected by one another, either through competition or symbiotic relationships. Often they compete with one another for nutrients or energy, much of which is derived from organic matter, but frequently organisms, having been nourished by organic matter, become an energy and nutrient source for other organisms. In many ways, the interdependence or the competitiveness of the various organisms affects soils and plants growing on them. Most of the activities of soil organisms are beneficial for crop plants. In fact, the fertility of a soil is often related to the number and diversity of the organisms it can support.

The reasons for this are quite complex, but some of them become apparent as we examine some of the beneficial processes conducted by certain organisms. Numbers of organisms are related to the production of sufficient quantities of beneficial substances. Diversity ensures that ample quantities of these substances will be produced. Diversity also helps guarantee that no single organism will dominate the soil, thus avoiding the elimination of some vital processes or possibly allowing a pest to decimate the crop.

Diversity is also essential for the orderly decomposition of OM, with its release of bound nutrient elements into available forms, and to the formation of SOM. The process begins with the maceration of OM by the larger soil fauna (animals in the ecosystem). The decomposition of the OM is aided by the high O_2 level in the soil due to the channels and burrows formed as fauna move about in the soil. These channels and burrows play a significant role in aiding drainage and aeration. The mixing of OM as these animals move about in the soil also aids decomposition, as soil rich in microorganisms is now in intimate contact with OM. It appears that the more resistant forms of macerated OM are first softened by fungi and then eaten by small animals such as mites, whose digestion may be aided by a microbial population in their gut. The undigested excreted material becomes a food source for other animals as they help transform OM into SOM.

Earthworms play an important role in this transformation as they change the excreta of several invertebrates into forms that are finally converted to SOM by microorganisms. Earthworms, along with the microscopic organisms (fungi, actinomycetes, and bacteria), tend to reduce the C:N ratio of OM, which is so essential for the production of SOM. Although earthworms can reduce the C:N ratio further than the microorganisms, the final conversion of OM to SOM depends on the microbiological population.

The numbers and activities of microorganisms vary tremendously in different soils. Both the numbers and activities are affected by the soil and the stimuli received from plants or other organisms. Many organisms are relatively inactive unless stimulated by root exudates or plant residues that are undergoing decomposition. Their presence and activities also are often interdependent, making their existence and importance highly variable at times and making them dependent upon cultural practices, which may benefit one group over another.

Much of the importance of soil microorganisms and their dependence on organic matter cannot be fully understood without taking due account of the rhizosphere, or the zone of soil that is influenced by plant roots. The zone varies somewhat for different plants and soils, but it usually extends less than 2 mm from the root. Despite the relatively small volume of soil involved, much of the microbial population thrives in this region. The populations increase probably because of a continuous deposition of plant cells that are sloughed off the root cap and epidermis as the root lengthens. The cells evidently provide valuable nutrients and act as an energy source for the different microorganisms present.

The major groups of organisms affecting organic matter are small animals, arthropods, earthworms, mollusks, nematodes, algae, protozoa, fungi, actinomycetes, and bacteria. It will be helpful to describe these organisms and some of their functions.

Small Vertebrate Animals

Several small animals (gophers, mice, moles, and squirrels) spend a good part of their life cycle in the soil. They feed mainly on recently dead plant materials, but are known to forage on living plants, often severely enough to become serious pests. Much of the damage to living plants seems to occur when there is a shortage of organic matter.

These animals have an effect on the soil and on formation of SOM. In their search for food and their nest building, they tend to open soils, leaving channels that greatly aid water and air movement. They tend to break down OM by shredding the raw product, which facilitates further decomposition by other organisms. Much of the macerated OM and the animal feces are distributed in a greater volume of soil as the animals move through it, thereby helping to enrich greater depths of the soil with SOM. Finally, when the organism dies, it is a source of energy for many other soil inhabitants.

Arthropods

The arthropods are a diverse group of organisms with jointed legs. The group includes various insects (beetles and their larvae, sowbugs, springtails, termites, ants), spiders, mites, and millipedes. They are found in most soils, but are present in great numbers in soils rich in organic matter. Forest soils, pastures, and no-tilled or minimally tilled soils usually have great numbers. Although some arthropods can cause damage to roots and thus spread certain root diseases, many feed on disease-causing pests. Others greatly improve soil porosity improving water intake, drainage, and aeration. Their droppings contribute to SOM and, in soils with low numbers of earthworms, the excreta could be an important part of humus. Their major contribution could well be shredding plant residues and mixing them in the soil, thereby stimulating decomposition of the residues by microorganisms.

Insects

Although all arthropods play a part, either in organic matter decomposition and humus formation or in the autonomy of soil organisms, it is the insects that receive the major attention, primarily because of their effects on crops. Many insects inhabit the soil, mainly in the upper 6 inches. Some of the factors that affect their kinds and numbers, which are highly variable, appear to be type of crop, climate, type of soil, and the amount of organic materials present. Some soilborne insects have value, as they mix soil by their movements. Termites, present in tropical zones, can move tremendous amounts of soil, at times forming mounds that interfere with cultivation. Movement of soil by burrowing insects is much less striking but still can be a means of aerating soil and aiding water movement.

Many insects overwinter or spend much of their juvenile development in the soil. Unfortunately, some larvae of these insects feed on the roots of many plants, debilitating them, often exposing the roots to infection by bacteria and fungi. The larvae of Japanese beetles, *Popillia japonica* (Newman), can be particularly damaging to a host of plants, but, fortunately, introduction of milky spore disease organisms has reduced damage from this source. The Diaprepes root weevil, which causes serious damage to roots of citrus, ornamentals, and sugarcane, still poses serious damage to these crops, although some insecticides have been introduced that show promise of control.

Organic matter may have a place in controlling some destructive larvae. For example, eggs and larvae of the Colorado potato beetle (*Leptinotarsa decemlineata*) were greatly reduced by predators in the presence of a wheat straw mulch placed soon after potato plants emerged. Nonmulched plots suffered 2.5 times more defoliation than mulched plots, which probably accounts for the mulched plots producing 25 to 32 percent more tubers than the nonmulched plots (Brust 1994). More about the use of mulches to control pests is presented in Chapter 8.

Despite the usefulness of organic matter in controlling some pests, neither OM nor SOM can be considered a universal practical control. In fact, some plant-damaging insects may be present in roots or stems of plants, and the residue (OM) allows the insects to survive until the next crop is planted. Control at times is possible by deep plowing that buries the insects and OM in an unfavorable environment or by burning, but these methods reduce the effectiveness of plant residues, and should be discouraged wherever possible in order to maximize the benefits of the residues (see Handling Crop Residues for Pest Control in Chapter 6). Some control is obtained as large amounts of OM and SOM promote the development and/or maintenance of a large and varied population of diverse creatures, many of which feed or limit the production of harmful insects, as in the Colorado potato beetle example.

Ants and termites also can be serious pests. Most of the time they are more of a problem for animals and humans rather than plants, although ants do move aphids from plant to plant. The aphids damage plants by both the removal of plant sap and transfer of some viruses. Ants, particularly the imported red fire ant *(Solenopsis richteri),* sometimes reduce yields of farm crops by feeding on seeds and seedlings, but most of their damage consists of stinging animals and people. At times their mounds, and especially those of termites, can interfere with cultivation of crops.

Earthworms

Earthworms are usually abundant in deep soils with ample calcium and adequate moisture holding capacity that are well drained. They do not do well in dry soils, but soil moisture cannot be excessive, as they need ample O_2 for respiration. A heavy rain or irrigation can force them to the surface, where they can dehydrate rather rapidly or become a meal for birds and other animals. They are most numerous in soils with surface residues that can be used as food and provide protection from dehydration and predation.

They feed on a variety of materials, many of them subsisting on the dead roots of grasses and clovers, and some on surface litter, which may be dragged down into their burrows. The great numbers of earthworms in soils rich in organic matter are considered as the principle mixing agent for moving dead surface litter into the soil where it may be attacked by soil microorganisms. Their burrowing increases porosity by leaving channels through which water and air move rapidly and roots can easily penetrate. While burrowing, earthworms ingest soil and organic matter and excrete casts that improve structure in many soils.

One of the benefits of earthworm activity is the deposition of C and N at the surface. The casts, which have a high resistance to raindrops, are rich in C and nutrients. Casts mixed with soil have increased corn growth and N uptake (Hauser et al. 1997). Leachates from casts have broken seed dormancy and promoted radical growth of jute (*Corchorus olitorius* L.) (Ayanlaja et al. 2001).

Mollusks

Slugs and snails are more numerous in moist soil, where they feed largely on dying vegetation, although they can become serious pests as they devour live plants when the supply of fallen leaves or old grass is diminished. Fortunately, such injury is not common on arable land, but their feeding can be a serious problem in gardens.

They have some value in making OM a more digestible food for microorganisms as they feed on dead OM. Also, at least one genus *(Testacella)* consumes worms, centipedes, and other slugs (Russell 1961).

Nematodes

Nematodes or eelworms can be present in large numbers in arable soils, ranging from 2 to 20 million per square meter. About 200 different species have been cataloged in some soils, although only about 20 of these are relatively common. The following three groups have been recognized: (1) a group that lives off soil microflora and possibly decaying organic matter; (2) those that live off soil fauna including bacteria, fungi, protozoa, and nematodes; and (3) those that are parasitic on plants (Russell 1961).

Because of its economic impact, the parasitic group has been studied more than the others. Nematodes are obligate specialized parasites that can reproduce normally only on certain kinds of plants. Such plants are susceptible, but susceptibility varies from a condition where most plants will be seriously injured to one in which most plants are not injured or seldom injured.

Fortunately, most nematodes are not parasitic. Unlike the parasites, the nonparasitic types are relatively mobile and depend on an open soil for free movement. Compaction tends to reduce the numbers of nonparasitic types as they require considerable movement to survive, but does little to affect the parasites that are relatively immobile. Many of these nonparasitic nematodes feed on parasitic nematodes as well as algae, bacteria, protozoa, and fungi. Also, mermithid nematodes can parasitize some insect species that are harmful for crop production. There is little evidence that they feed on plant residues. Plant residues and other forms of organic matter can, however, reduce the number of parasitic nematodes, probably by supporting other organisms that may compete with or feed on the parasitic types.

Algae

Algae are microscopic forms that contain chlorophyll and are usually found at the soil surface or in the upper few inches of soil. They are most numerous (generally 100,000-200,000 per g of soil) in damp soils that are not exposed to excessive sunlight which may dehydrate them. Greater numbers usually exist on finer-textured soils with ample available N and P as compared to coarser soils with less fertility. Some algae are capable of fixing N from the atmosphere. The optimum pH for such fixation appears to be in a slightly alkaline range of 7-8.5, with fixation taking place in the range of 6-9 (Russell 1961).

Nitrogen fixation by algae probably is of little importance in most arable soils because frequent tillage appears to disrupt the colonies close to the surface, but may be appreciable in minimally tilled soils that maintain moist surfaces. There is some evidence that appreciable amounts of N are fixed under dense cover crops, sugarcane, and grassland, where surface soil may be moist a good part of the time. Nitrogen fixation probably is greatest in rice culture, which has wet soils for much of the year, although alternate drying and flooding commonly used in some rice culture appears to be essential for N fixation by algae.

Protozoa

Protozoa are minute animals that inhabit most of the soils in the world. Their numbers in some English soils appear to be greater in spring and autumn than in summer and winter. Average numbers per g of soil taken from long-term plots at Rothamsted were 17,000 for unmanured plots, 48,000 for plots receiving complete fertilizer, and 72,000 for plots receiving manure. The size varies from about 5 to 20 µm in length for flagellates (those which have flagellae to help them move), 10 µm to several tenths of a mm for amoebas, and 20 to 80 µm for ciliates. Generally, protozoa are restricted to soil pores that contain water in which they can move about, but they are capable of forming cysts, which permit them to survive for long dry periods and become active again with the first rains (Russell 1961).

Most protozoa feed largely on bacteria, but some algae, yeasts, and amoebas may be consumed. A few contain chlorophyll and can produce their own sources of energy, while others can take nutrients directly from solutions. Once thought to be highly detrimental to all bacteria, it is now known that certain bacteria are consumed in great numbers while others are seldom consumed or not at all. The inedible bacteria or those seldom consumed appear to excrete substances that are toxic to amoebas, providing a selective influence on the bacterial population. Furthermore, protozoa, instead of being harmful to all bacteria, seem to enhance the efficiency of certain bacteria, such as *Azotobacter*, which fix atmospheric N into forms suitable for plant use. Amoebas also appear to influence bacterial efficiency by affecting the oxidation rate of sugars (Russell 1961).

Fungi

Fungi found in soils are small plants that vary tremendously in size from microscopic, single-celled organisms, not visible to the naked eye, to mushrooms that may be more than several inches in height and width. They are very numerous in many soils, consisting of many thousands of species. Only bacteria are present in greater numbers, but the fungi provide a greater mass because of their relatively greater size. They can grow in soils of wide pH variation but often become dominant at low pH values (4.0-5.5) where bacterial populations may be limited. Generally, fungi are aerobic, but some of them tolerate wet or compacted conditions with relatively low O_2 and high CO_2 concentrations.

The fungi appear to have slower turnover rates than bacteria, probably accounting for greater nutrient retention and the enhancement of SOM formation in fungal-dominated soils as compared to those dominated by bacteria. They process a wider range of compounds than most bacteria or actinomycetes, metabolizing starch, cellulose, gums, and lignins as well as sugars and proteins. In so doing, they greatly aid in formation of SOM, often completing the breakdown of OM initially started by other organisms.

Some fungi, primarily molds and mushrooms, have hyphae, which are thin long strands often twisted together to form mycelia. These are readily visible running through decaying OM. The molds minus their mycelia are generally microscopic. The penetration of decaying matter not only helps in OM breakdown but is important in stabilizing peds formed as substances released in OM decomposition cement the various particles together.

Many fungi play an important role in the decomposition of organic matter and the temporary immobilization of N. Nearly all soil fungi depend on extraneous N, which after uptake is released rather slowly. The saprophytes, probably because of their rapid growth in the presence of an ample food supply, can compete with crop plants for available N. The problem is evidently accentuated in no-till residue or residues that have a wide C:N ratio. Recent studies indicate that an application of the fungicide Captan can reduce decomposition rates of surface-applied residues by 21-36 percent. Immobilized N in soil not treated with Captan was fourfold greater than that of treated plots. These studies indicate that fungi play an important role in decomposition of organic matter, but by building N into mycelia, fungi regulate the uptake of N and its availability for crop plants (Beare 1997).

Most fungi cannot fix N and are not capable of producing their own food by photosynthesis, relying upon other sources for their energy. The saprophytes depend on dead plants and animals: the parasites gain their sustenance by invading living plants and animals, and the symbionts fill their needs by symbiotic relationships with a number of plants.

Saprophytes

The food sources of saprophytes vary. Some can digest only sugars or simple carbohydrates from organic matter that is easily decomposed, while others are able to digest cellulose but not lignins, and others can utilize lignins as a source of energy. The first group depends on dying or dead tissue that is quickly invaded by mycelia. The cellulose decomposers grow more slowly, and those that can digest lignin are the slowest growers.

The saprophytes play an important role in SOM production because of their large mass, their slow decay, and the ability of some of them to synthesize humiclike substances.

Parasites

The parasitic fungi can cause serious losses of crop plants. Some of the more common diseases are listed under Disease Suppression. Once introduced, many parasitic fungi can exist for a number of years because of the formation of resistant spores. These spores can resist adverse conditions and germinate when conditions of moisture and temperature are ideal.

Conditions favoring parasitic fungi are variable but involve the presence of a host plant. Some parasitic fungi are crop specific or at least species specific, but others can seriously damage several different kinds of plants. Some can be deadly to a particular plant but survive in other plants with much less damage to them. Continuous planting of the host plant usually encourages large increases in the parasite, although there have been reports that at times continued monoculture of a host plant will lead to disease-suppressive soils that prevent the development of the disease. Exactly why this happens is not fully understood but is thought to be due to induced resistance as some nonpathogenic fungi invade the host.

Some parasitic fungi are not always parasitic but cause serious damage only as conditions are very favorable for them or very unfavorable for the host plant. Some fungi may be saprophytic but become parasitic under certain conditions. Poor plants due to poor nutrition or unfavorable soil physical conditions are particularly prone to attack by these and more virulent fungi. Many fungi cause serious trouble when soils are wet and/or cold. Some fungi, such as *Fusarium* spp. and *Pythium* spp., can thrive on lower O_2 levels than their host plants. Overly wet or compacted soils usually lead to much more Pythium damping off or Fusarium wilt. The importance of porosity for improving soil O_2 and its dependence on organic matter is presented in Chapter 4.

Symbiotic Forms

As presented in Chapter 3, a mycorrhiza is a group of fungi that invade plant roots, but form a symbiotic (mutually beneficial) relationship that greatly aids plant nutrition. The relationship between mycorrhizae and many plants in natural ecosystems probably ensures their survival, as many of these plants likely could not survive without them. The relationship is also advantageous to cultivated plants, but its full potential seldom is reached due to certain fertilizer and pesticide practices.

As is pointed out in Chapter 3, mycorrhizae extract sugars from plant roots for their energy source, but plants gain some real advantages in return for this small loss of sugars. Mycorrhizae help protect the roots from certain soilborne disease organisms and parasitic nematodes, and generally tend to increase growth in the host plant.

An added advantage of mycorrhizae comes from their production of glomalin, which greatly aids soil aggregate formation. The glomalin appears to glue together various particles of organic matter, consisting of plant cells, fungi, bacteria, and microorganisms, with soil particles to form large peds. As pointed out in Chapter 4, formation of peds and their building into larger units are needed for good porosity, infiltration, and drainage.

With all of the advantages of mycorrhizae, there appear to be several constraints on their efficient use for cultivated crops, or there would be a better utilization of their value. They are adversely affected by the application of certain insecticides and herbicides. Very low or high levels of soil nutrients seem to reduce their effectiveness. Also, it appears that mycorrhizae may become parasitic under extreme cases of very low levels of soil nutrients. The opposite condition of ample nutrients, especially a good supply of P, is more common in reducing their effectiveness. Generally, mycorrhizae are most efficient in soils of relatively low fertility that receive little fertilizer, although some application of N can be stimulatory. On the other hand, they are usually more significant in soils of ample organic matter, where crops are rotated and cover crops are included but receive minimum or no tillage.

Actinomycetes

Actinomycetes are unicellular organisms that form a transition between fungi and bacteria. They are very numerous in many soils, often present in greater numbers than any other soil organism except bacteria. They are especially numerous in soils rich in organic matter such as sods, but are inhibited by pH values less than about 6.0. Most of them depend on decaying organic matter for their energy source, but some of them are capable of forming a symbiotic relationship with certain trees, obtaining sugars from the tree but fixing N in return.

They play an important role in the decomposition of organic matter, utilizing the more resistant compounds, and often becoming the dominant species once the simple compounds (sugars and amino acids) are taken up by many different organisms.

In addition to the decomposition of organic matter with its important release of nutrients, actinomycetes play an important role in the development of antibiotics. Actinomycetes grown in pure cultures have been used to produce a host of antibiotics, such as streptomycin, neomycin, and agrimycin, which have been most effective in fighting human and animal diseases. Their production of antibiotics under natural soil conditions probably aids their survival, but also must be of considerable help to crops by preventing dominance of certain soilborne disease organisms.

Bacteria

Bacteria are small single-celled organisms of many different types that are present in great numbers in most soils, ranging from millions to billions and even trillions per g of soil. Their numbers vary depending on organic materials present and will increase rapidly with the addition of OM or as soils are moistened after a drought or warmed in the spring. They can form resting stages capable of surviving long periods of adverse conditions including low food supplies, but will revive as conditions once more become favorable. Their rapid reproduction rate allows a very fast renewal—sometimes within hours after conditions become favorable.

There is a great variation in the nutritional requirements for energy and C sources among different bacteria in the soil. Most, considered heterotrophs or organisms that derive their energy only from decomposition of organic matter, seem to prefer easily decomposable materials, such as sugars, starches, amino acids, or simple proteins. Partially decomposed materials present in animal manure or composts are excellent sources of energy and C for these bacteria. Others, designated autotrophs, derive their energy from the oxidation of H, NH_4-N, NO_2-N, Fe, or S and obtain their C from CO_2 or various carbonates. A few bacteria, existing primarily in wet soils, contain chlorophyll and obtain their energy through the photosynthetic process.

The differences are reflected in their mode of survival. Like the fungi, bacteria can exist with different food sources. Most are saprophytes living off decaying OM. They are responsible for most of the OM breakdown but primarily use the simpler compounds of sugar, starches, and amino acids, leaving the more complex and more resistant compounds of cellulose, lignins, and chitins to the fungi.

A small group is parasitic, although some of these are not always parasitic. They may be saprophytic most of the time, but become parasitic as conditions are either favorable for the bacteria or unfavorable for the plant. Often parasitism occurs because the bacterium can thrive at either higher temperatures and/or with more moisture than the plant. Soft rot of carrots and rotting of potato seed pieces caused by *Erwinia carotovora ssp. carotovora* is an example of a bacterial disease brought on by excessive moisture and elevated temperatures. Both rots cause serious losses if extended irrigation allows a film of water to remain on the carrot roots or potato seed pieces and if temperature is elevated. More about soil bacterial diseases is presented under Disease Suppression.

A very important group of bacteria are symbionts that form symbiotic relationships with plants. The most important symbiotic relationship involves the fixation of atmospheric N_2, which is unavailable to plants, into an available form. The relationship, presented in Chapter 3, is covered in more detail under Nutrients Gained.

Most bacteria do best in well-aerated soils having near-neutral pH values with ample Ca. Evidently due to the extreme variability among different bacteria, some will be found growing over a wide range of moisture and pH values. Facultative bacteria, or those capable of growing under both aerobic or anaerobic conditions, will continue growing under wet conditions, evidently because they can reduce chemicals other than O_2 for respiration purposes. The reduction of N compounds under wet conditions results in loss of N by volatilization.

The various forms of bacteria also allow for both oxidation and reduction of several elements in soils, which can have substantial effects on plant nutrition. Some aspects of oxidation and reduction on plant nutrition were covered in Chapter 3.

PROCESSES VITAL TO SOIL HEALTH

Several soil processes, carried out by soil organisms and dependent on organic matter, are essential for soil maintenance and the sustainability of agriculture. Some of the more important processes are (1) nutrient cycling, (2) the increase in available nutrients brought about by N fixation and the activities of mycorrhizae, (3) the improvement of soil physical properties by formation of aggregates, (4) the detoxification of harmful substances, and (5) maintenance of balanced ecosystems.

Nutrient Cycling

The uptake of nutrients by various organisms, including plants, and the release of these nutrients as the organism decays provide an orderly supply for plants over a period of time. As has been indicated, small vertebrate animals, arthropods, insects, earthworms, and nematodes initiate the decomposition of organic matter, but bacteria, along with fungi and actinomycetes, are responsible for the final breakdown of plant and animal residues. The process releases the nutrient elements contained in various organic compounds in forms that are readily usable once again by green plants. Not all of the nutrients present in organic matter are immediately available for plant use as soon as the organic matter undergoes decomposition. A large portion of these nutrients may be taken up by the organisms involved in the process of decomposition, becoming available for plants only after these organisms die and decompose.

A cycle of release and uptake may extend for a long period, which has several advantages. It allows for a much more orderly plant growth than if all of the nutrients were released in a very short period. There is much less chance of nutrients being lost by leaching. The long period of release permits a number of organisms to prosper in the soil and, in so doing, improve the physical nature of the soil, for example increasing porosity, improving water permeability and drainage, and allowing for better gas exchange.

It is possible for all the elements essential for plant growth to be present in plant material, but at times one or more elements may not be present in amounts high enough to promote plant growth. Plants produced on soils poor in required elements usually contain insufficient quantities of these elements to adequately support crops, making it necessary to add the missing elements by fertilizers or organic materials. Soils that have been farmed for long periods of time or that have been exposed to intensive leaching without fully replacing elements removed by crops are very prone to producing organic materials inadequately supplied with the needed elements. Not only is the quality of the organic materials in question, but the amounts produced are also often inadequate, making depletion of SOM a foregone conclusion.

Nutrients Gained

In addition to the nutrients released from organic matter as it decomposes, substantial amounts of nutrients are added to the soil or made available to plants by several biological processes that are aided by organic matter.

The biological processes include the symbiotic and nonsymbiotic relationships between plants and microorganisms, oxidation-reduction reactions initiated by microorganisms, aggregate formation which favors nutrient enhancement by providing a ready exchange of gases that assures adequate respiration, and the detoxification of harmful substances that may interfere with nutrient availability. These important processes adding to the quantity of available elements are described in the following sections.

Nitrogen Fixation

The presence of organic matter enhances the fixation of atmospheric nitrogen (N_2) from a form unavailable to plants to one that is readily utilized. The fixation of N is largely accomplished by bacteria, although some N is fixed by actinomycetes and a very small amount by algae. As has been pointed out, fixation by bacteria can either be symbiotic or independent. The symbiotic organisms are not directly dependent on organic materials for their source of energy, as they receive ample amounts from their hosts, supplying the host with valuable quantities of N in return. The nonsymbiotic organisms depend on organic materials as their source of energy for their maintenance as well as the fixation of N.

Symbiotic nitrogen fixation. Although symbiotic organisms are not directly dependent on organic matter as a source of energy, organic matter does play an important role in the symbiotic fixation of N by providing soil conditions very favorable for the host plant. In fact, active nitrogen fixation takes place only if the host plant is well supplied with mineral elements necessary for growth, and organic matter can aid in providing a sufficient supply. Also, organic matter is helpful in supplying sufficient soil O_2 for the increased respiration of plants supporting an abundant population of symbiotic organisms in root nodules by altering the porosity of soils to favor drainage and exchange of gases.

In the most prominent and studied symbiotic relationship, *Rhizobia* bacteria, living in nodules located on legume roots, transform atmospheric N that is unavailable to plants into a usable form. In the association, the rhizobia receive sugars from the plant. The sugars are a source of energy for the bacteria and for transforming the relatively inert N_2 of the atmosphere to ammonium nitrogen (NH_4-N), which can be utilized by the host plants. A similar relationship exists between *Bradyrhizobium* spp. and some legumes, such as cowpea, peanuts, pigeon peas, and several other tropical legumes. Both of these organisms form nodules on the roots of their hosts, but some bacteria, such as *Azotobacter* and *Azospirillum* spp., fix N in association with several grasses but do not form nodules.

In addition to nutrient supply and pH value, the effectiveness of the fixation depends largely on the strain of organism infecting the plant. Certain species or subgroup strains of rhizobium bacteria may be effective only on one kind of legume, although some may be effective on several different kinds. For example, *Rhizobium leguminosorum* bv.*viceae* does an effective job with peas, sweet peas, vetch, and lentils, while the subgroup bv. *trifolii* is much better adapted for inoculating various clovers, but the species *R. fredii* is almost limited to soybeans.

Appreciably large amounts of N can be fixed by the symbiotic process, but amounts vary from about 25 to over 400 lb/acre. The amounts produced vary with various plants and soils and with the organisms responsible for the fixation.

Some strains of bacteria are relatively inefficient, fixing small quantities of N. Bacteria symbiotic with beans and peas fix such small quantities of N that these crops need to receive additional N to produce valuable crops. Generally, organisms producing nodules on the roots fix greater quantities of N than the nonnodulated types. The nodules apparently protect the nitrogenase enzyme system that is responsible for the reduction of N_2 to a form available to the plant.

The N fixed by symbiosis benefits the organism and the host plant, but it may also benefit other nearby plants as well. Often, nonfixing plants associated with the host plant receive enough N from the process to benefit from it. The better yields of grass, when combined with clovers in pastures, is a common example of such benefits. Evidently, legume roots and nodule tissue release considerable N as these tissues break down.

Whether soil N will be improved by the symbiotic fixation of N depends somewhat on the amounts fixed and how the host crop is handled. Soil N usually is increased from alfalfa grown for a few years or from incorporating the entire crop of annual legumes, such as hairy vetch, but may show no increase if seed is harvested.

Nonsymbiotic fixation of nitrogen. Nonsymbiotic N fixation is largely accomplished by bacteria, although some atmospheric N is fixed by actinomycetes and a very small amount by algae. Organic matter supplies the energy for the microorganisms responsible for nonsymbiotic fixation. Nitrogen is made available for any plant, the roots of which are in close proximity to the organisms, although the amounts produced vary with different soils and with the organisms responsible for the fixation. The amounts are generally much smaller than those produced by symbiotic organisms.

The most studied nonsymbiotic organisms capable of fixing N are bacteria belonging to the genuses *Azotobacter, Beijerinckia,* or *Clostridium.* The *Azotobacter* do best in soils that have a pH of at least 6.0, a good supply of P, and good aeration. The *Beijerinckia* are more acid-tolerant and can do well with less Ca than the *Azotobacter.*

Clostridium bacteria are found in many different soils and probably can fix N in soils of lower O_2 content than the other two genera. It is believed that all of these organisms use the same process of fixing N, which depends on an enzyme reaction that requires a supply of Mo and is suppressed by CO or H_2 (Russell 1961).

Mycorrhizae and Available Nutrients

As was indicated in Chapter 3, mycorrhizae improve the availability and/or uptake of several important nutrient elements for crop plants. The improvement in availability for sorghum plants infected with VAM (vesicular-arbuscular mycorrhizae, contain both vesicles and arbuscules that store and help transfer nutrients) is indicated by the higher contents of P, K, Ca, Mg, S, Cu, Fe, Mn, and Zn present in infected plants compared to noninfected plants. The shoot and root dry matter of the infected plants were also much greater than in plants without mycorrhizae (Raju et al. 1987).

The improved uptake of nutrients by plants infected with mycorrhizae has been largely attributed to an extended root system created by mycorrhizae hyphae and to the ability of the hyphae to extract elements not normally available to plant roots. It is perhaps the effect of extracting nutrient elements from nonavailable sources and keeping P available that makes the uptake of P by mycorrhizae-infected plants so interesting.

In soils of low P fertility, VAM mycorrhizae are able to utilize P sources not normally available to plant roots. Mycorrhizae probably cannot effectively use P in highly insoluble sources, but there are indications that low or moderately soluble P sources are readily utilized. Mycorrhizae accumulate P from these and soluble sources of P at a much faster rate than root tissue, evidently because they are capable of storing it in an available form in the sheath until the plant requires the P. By so doing, the mycorrhizae act as a reservoir of available P for the plant, permitting continuous growth even if soil conditions may not provide sufficient P for noninfected plants (Tinker 1980).

The infection of plant roots with mycorrhizae takes place more readily under conditions of low P availability. In fact, infection occurs whenever the inflow or diffusive transfer of P to unit length of root is appreciably less than the amount of P necessary to maintain plant growth (Tinker 1980). Plants grown on soils fertilized with large quantities of available P usually are poor candidates for mycorrhizal infection. Fertilization, especially large doses, generally inhibits mycorrhizal infection. Application of other intensive inputs such as herbicides, fungicides, and insecticides can also deter the infection. On the other hand, procedures that enhance the accumulation or effectiveness of organic matter, such as reduced tillage, tend to favor the mycorrhizal infection.

Oxidation-Reduction Reactions

The oxidation of elements by bacteria takes place in well-aerated soils, while reductions occur in soils with low O_2 contents, such as waterlogged or compacted soils. Several of the oxidation-reduction systems are reversible, shifting to the reduced form when O_2 is in short supply or back to the oxidized from as O_2 becomes more plentiful. Typical inorganic oxidation-reduction systems in soil are sulfate-sulfide (SO_4^{2-}-S^{2-}), manganic-manganous ions (Mn^{3+}-Mn^{2+}), and ferric-ferrous ions (Fe^{3+}-Fe^{2+}). The oxidation-reduction of carbon (C^{4+}-C^{4-}) also takes place on a regular basis. The oxidation of NH_4-N-NO_3-N readily takes place in well-aerated soils, and the reduction of nitrate-nitrite (NO_3^--NO_2^-) readily takes in wet soils.

Oxidation and reduction reactions can be favorable or unfavorable for plant growth depending upon soil conditions and type of plants. For example, the oxidation of C to CO_2 taking place in most well-aerated soils amply supplied with organic matter is beneficial or benign for most conditions. This oxidation releases energy for microorganisms and plants, and the CO_2 formed can readily be replaced by air with ample O_2, allowing continued respiration in the roots. Oxidation of C is preferred to its reduction, which can take place in overly wet or compact soils usually low in organic matter. Reduction of C can produce methane that is toxic to a number of plants.

The oxidation of S that produces the sulfate ion (SO_4^{2-}) is beneficial for plants. Plants cannot absorb S, but readily absorb and utilize the sulfate ion (SO_4^{2-}). If soils lack sufficient O_2, S can be reduced to hydrogen sulfide (H_2S), which can be lost to the atmosphere or can be potentially damaging to plants if it remains in the soil.

In most soils, the reduced forms of iron and manganese (Fe^{2+}, Mn^{2+}) are preferred to the oxidized forms (Fe^{3+}, Mn^{3+}), as the reduced forms are more readily available to plants. But in some soils, usually those with low pH values, the added amounts of reduced forms may be toxic to a number of plants. The additional reduction of these elements often causes or aggravates toxicity by increasing the amounts available. The amounts of Fe and Mn in acid soils are usually already sufficient as a result of the normal increased solubility at lower pH values and in some cases due to reduction by bacteria under anaerobic conditions. Organic matter can be useful in reducing problems of excess Fe and Mn by limiting the drop in pH because of its buffering effect and by providing better soil aeration, thus limiting the amount of reduction.

Formation of Aggregates

As has been pointed out in Chapter 4, the production of glues capable of cementing soil particles into larger entities or peds is largely a result of bacterial and fungal activities. Exudates from plant roots also are beneficial. The glues or mucilages produced by bacteria, fungi, and actinomycetes help cement the individual soil particles into peds and build the peds into macroaggregates. Without sufficient macroaggregates, it is difficult to obtain adequate water infiltration or drainage and enough gas exchange to remove CO_2 and introduce enough O_2 necessary for plant and microbial growth.

Detoxification of Harmful Substances

A number of substances capable of affecting plant growth adversely are formed in the soil or are applied to the land as fertilizers, pesticides, composts, biosolids, or deposited as pollutants from industrial activities. The continual detoxification of these substances is of prime importance lest they accumulate in concentrations high enough to seriously limit crop production.

Soil bacteria and fungi are the major organisms involved in this soil remediation. Composted biosolids that are rich in these microorganisms are useful in establishing turfgrasses on disturbed urban soils (Loschinkohl and Boehm 2001). Some of the microorganisms thrive on heavy metals or specific inorganic or organic compounds that can cause problems, evidently by producing enzymes that can break down the toxic substances so they can be used as foods. Although their numbers may be small in most soils, these organisms seem to increase in the presence of some toxic substances.

CONTROLLING PESTS BY HARNESSING SOIL ORGANISMS AND ORGANIC MATTER

Organic matter, both as OM and SOM, is involved in the suppression of various plant pests. As discussed previously, the suppression can be the result of indirect effects upon the soil environment that favor the plant or due to a large, diverse population of soil organisms. The improved SOM resulting from the additions of various forms of OM and its proper placement and the increase in biota that follows opens the possibility of controlling pests by the judicious use of organic matter.

Balanced Ecosystems and Pest Suppression

There is a general consensus that a balanced ecosystem is beneficial in suppressing a wide variety of soil pests, and organic matter helps maintain a balanced ecosystem in the soil. Organic matter, by supplying nutrients and energy as well as making the soil a suitable habitat, allows the various soil inhabitants described in the first part of this chapter to thrive in great numbers. While many survive, there is little chance for dominance by any one group, due to the competition between them. As a result, it is possible for the various beneficial processes vital for soil health to be carried out by the different organisms, and still control several plant pests.

Usually, the organisms causing plant disease (bacteria, fungi, and nematodes) are considered by plant pathologists without paying due attention to many more nonpathogenic organisms existing in soils. For example, there have been estimates of 1,000,000 to 10,000,000 free-living nematodes, 10,000 to 100,000 enchytraeids (pot worms), 1,000 to 10,000 mollusks (slugs and snails), 100 to 1,000 myriapods (millipedes and centipedes), 100 isopods (wood lice), 100 Araneidae (spiders), 1,000 Collembola (springtails), and 10,000 Acranaria (mites) per square meter of soil (Campbell and Neher 1996).

They all are involved in the decomposition of organic matter and affect nutrient cycling in the soil. Nematodes and protozoa affect the amounts of available nutrients by excreting nitrogenous wastes, primarily as ammonium ions. Other microinvertebrates increase soil fertility through the deposition of feces and acting as a reservoir of nutrients that are released as they die. Various organisms move bacteria, fungi, and protozoa by ingesting them or carrying them on their cuticle, and in so doing help to colonize organic matter, which aids in decomposition.

Mechanisms of Pest Suppression

The balance between various soil organisms is a result of (1) competition for food and energy, (2) substances produced by one organism that are harmful to another (antagonism), (3) the ingestion of one organism by another, and (4) parasitism of one organism by another.

Competition for food and energy. The part played by the competition for food and energy between various soilborne organisms in pest control has been touched upon. A large supply of organic matter ensures that great numbers of different types of soil organisms develop and survive, but competition for food and energy limits the numbers of any one group.

Antagonism. One of the benefits of a balanced ecosystem is that it often includes a number of microorganisms that are antagonistic to disease organisms. The antagonism is expressed through competition for food and energy, parasitism of the disease organism, feeding on the disease organism, production of substances that are harmful to the disease organism, or by inducing disease resistance in the host plant.

Substances produced by soil organisms that are harmful to pests have received considerable attention, and some have been put to use to control plant, animal, and human diseases. Most of these substances fall under the designation of (1) antibiotics and (2) bacteriophages.

1. For years, soil microbiologists were intrigued with two phenomena as they tried to grow microorganisms in artificial cultures. One dealt with the inability to grow the organism in the medium despite the presence of an ample food supply. The other dealt with the clear zones often exhibited around certain microorganisms as they grew in petri plates. Evidently, these organisms were producing substances that definitely inhibited the growth of their neighbors. The clear zones were especially interesting since they influenced the growth of

neighboring organisms and offered a chance to isolate the material(s) that offered promise for the control of microorganisms. We now know that the phenomena are due to the production of antibiotics. The first, which tends to "stale" the development, is largely due to penicillin, produced by the fungus *Penicillium notatum.* The second is primarily produced by a group of actinomycetes, which are responsible for the production in pure cultures of the commercial antibiotics actinomycin, neomycin, and streptomycin.

Dr. Selman Waxman and his co-workers at Rutgers University in the 1930s isolated some of the substances producing the clear zones later managed to grow the organisms in commercial quantities in pure cultures. The production of streptomycin by the actinomycete *Streptomyces griseus,* the most famous of the antibiotics isolated by the group, has been effective for controlling tuberculosis in humans and several animal and plant diseases. Its use against tuberculosis is being phased out because disease organisms have become resistant to it. Resistance has also been the reason for its reduced use against plant diseases, except for fire blight caused by *Erwinia amylovora* Burr. There are doubts as to how long it may be effective against fire blight for the same reason.

Although resistance by disease organisms to repeated doses of antibiotics has reduced their value to control plant diseases, antibiotics produced in soils continue to inhibit the growth of a number of organisms and help maintain a healthy soil microflora balance. In fact, the presence of some antibiotics improves the suppression of soilborne pathogens by certain suppressive soils, and "superior agents" for controlling diseases could readily be derived from genetically engineered strains of microorganisms producing antibiotics (Thomashow and Weller 1996).

We should not overlook the effectiveness of fungi that can produce antibiotics. Those that produce antibiotics often have hyphae that are resistant to antibiotics unless present in large amounts. The fungus *Trichoderma virile* protects the roots of many plants because it appears to produce antibiotics, allowing it to grow over the mycelia of other fungi inhabiting the rhizosphere that are capable of causing damage to the root. *Trichoderma lignorum* evidently produces substances that are harmful to *Rhizoctonia solani,* which causes several serious plant diseases. Factors that favor the *Trichoderma,* such as soil acidification, often are sufficient to control several fungal diseases such as root rot of strawberries and tobacco caused by weak root parasitic fungi (Russell 1961).

2. Bacteriophages attack many different kinds of bacteria and a few species of actinomycetes, causing a lysis or dissolution of the outer membrane and loss of cell contents. They differ from antibiotics in that bacteriophages multiply in the living bacteria or actinomycete cells before lysis. The beneficial effects of bacteriophages in keeping a balance between organisms in the soil are often offset by their harmful effects to nodule bacteria (Russell 1961).

Ingestion of one organism by another. The ingestion of one organism by another appears to be a very important part of competition between organisms for food and energy, and it may be useful to cite a few examples. Although most slugs and snails live off dead and live vegetation, some, belonging to the genus *Testacella*, are predaceous on other slugs, worms, and centipedes. Springtails survive by ingesting not only decaying plant tissue but also the various microorganisms that decompose organic matter.

Earthworms depress fungal populations, probably by feeding on their mycelia. Nematodes feed on bacteria, algae, fungi, protozoa, other nematodes, and insect larvae. Mites feed on fungi and other microorganisms involved in organic matter decomposition. Protozoa primarily consume bacteria but also partake of some small algae, yeasts, and amoebas. Some fungi can attack small vertebrate animals, while others are harmful to nematodes or keep insects and mites under control.

Parasitism. Parasitism of one organism by another is a common occurrence in soils, not only of disease organisms but also of nonparasites. A few typical examples of parasitizing disease organisms can be helpful in recognizing their importance. *Penicillium* spp. appear to be the dominant parasite of *Sclerotium rolfsii* in composted grape pomace. The parasitism of several fungal disease organisms by *Trichoderma* and *Gliocladium* spp. is relatively common. In composts prepared from lignocellulose wastes, *T. hamatum* and *T. harzianum* interacting with bacterial isolates control Rhizoctonia damping off by parasitizing the causal agent. *Trichoderma* isolates also are effective in parasitizing *R. solani* but only in mature, not immature, compost (Hoitink et al. 1996). *Coniothyrium miniitans* is capable of parasitizing *Sclerotinia* and *Sclerotium* spp. *Sporidesmium sclero-*

tivorum is a parasite of *S. sclerotium, S. minor, Sclerotium cepivorum,* and *Botrytis cinerea* (Baker and Paulitz 1996).

Parasitism of soil insects also can help suppress their activities. One of the more striking examples is the control of Japanese beetle larvae by spores of two bacteria, *Bacillus popilliae* and *B. lentimorbus.*

Pests can be suppressed by practices that favor the survival of numerous soil organisms while preventing any one species from overwhelming the system. Some of these practices are: (1) handling crop residues for better pest control, (2) use of rotations, (3) increased use of cover crops, and (4) promotion of pest antagonists. (See Chapter 6 for discussion of the first three practices.)

Using Organic Matter for Specific Pest Control

Disease

Some of the more common soilborne diseases are wilts caused by *Fusarium* spp. or *Verticillium* spp., bulb rot of onions, banana wilt caused by *Fusarium* spp., southern stem rot or Sclerotium stem rot of peanuts caused by *Sclerotium rolfsii,* black shank of tobacco caused by *Phytophthora parastica* var. *nicotiana,* black root rot of tobacco caused by *Thielaviopsis basicola* Zopf, damping off of many plants caused by *Pythium* spp. or *Rhizoctonia* spp., corn stalk rots caused by *Diplodia, Fusarium,* or *Gibbela* spp., "take all" of wheat and other small grains caused by *Gaeumanomyce graminis* and diseases of many plants caused by *Thielaviopsis* spp.

Typical examples of soilborne bacteria causing plant diseases are *Agrobacterium tumefaciens* causing crown gall on certain fruits and nursery-grown euonymous plants and *Erwinia carotovora* causing soft rot. Other bacteria causing serious plant diseases are not necessarily soilborne but may survive on plant parts deposited on the soil. These include bacterial canker of stone fruits, angular leaf spots of cucurbits, bean halo blight, and bacterial wilt of carnations, all caused by *Pseudomonas* spp. Potato scab is caused by an actinomycete, *Streptomyces scabies.*

Organic matter can play an important role in controlling plant disease by ensuring a diverse soil population. Advantages of such a population have been outlined under Balanced Ecosystems and Pest

Suppression, but a few more specific examples of disease control by diverse soil inhabitants may be useful. The value of the fungus *Trichoderma lignorum* in suppressing *Rhizoctonia solani*, the agent causing many different diseases of a wide variety of plants, has already been mentioned. Also worthy of mention are (1) the grazing of *Rhizoctonia solani* by Collembolas, *Proisotoma minuta*, and *Onychiurus encarpatus;* (2) the perforations of *Cochliobolus sativas*, a fungus causing root rot of barley by giant amoebas; (3) the preferential feeding of Orabatid mites on pigmented fungi, such as *Rhizoctonia solani* and *Cochliobolus* over nonpigmented fungi; and (4) the preferred feeding of *Bradysia coprophila* on the sclerotia of *Sclerotinia sclerotiorum*, the organism responsible for lettuce drop on certain muck soils (Campbell and Neher 1996).

In two to three years in the Georgia program of rotations, intensified use of cover crops and reduced tillage have about eliminated damping off of cotton, peanuts, and vegetables due to attacks of *Rhizoctonia solani*, *Pythium myriotylum*, *P. phanidermatum*, *P. irregulare*, and *Sclerotium rolfsii*.

It needs to be pointed out that additions of organic soil amendments do not always tend to suppress disease, but at times may actually favor an increase. The exact causes for this phenomenon are not well known, but it has been speculated that some of these reversals are due to changes in the soil environment. A large amount of added carbohydrates can spur microbial growth so that large quantities of O_2 are consumed, while the production of CO_2 is greatly increased. The confluence of events coupled with restricted air movement, which may occur with the addition of large masses of wet material, may produce an O_2 shortage for the crop plant and an advantage for several disease organisms.

Recently, studies with composts indicate that the nutritional status of the compost determines whether the material favors disease suppression, although other factors, such as allelopathic toxins, C:N ratio, and concentration of soluble salts, especially Cl, affect the disease suppression of OM. In these studies, well-cured compost low in cellulose restricted germination of *Pythium* and *Phytophthora* spp., but immature compost rich in cellulose reduced the effectiveness of *Trichoderma hamatum* to parasitize *Rhizoctonia solani*, allowing the disease to take over (Baker and Paulitz 1996).

Induction of disease resistance. Induced resistance of crops to a disease by another organism is one of the ways in which soil organisms help to control plant disease. The mechanism of this phenomenon is not fully understood, but as a nonparasitic or an avirulent strain of the pathogen is introduced, several resistance actions are started, a common one probably being a signal that the host may be attacked by a pathogen.

It appears as if the host plant must be challenged by a potential pathogen or nonpathogenic isolate to induce resistance to the disease organism, although other inducers of resistance, such as injury, heat shock, or exposure to ethylene or degradation products of host and pathogen cell walls, have been noted. Inoculating cucumber roots with parts of a nonpathogenic isolate of *Fusarium oxysporum* induced resistance to *Colletotrichum lagenarium,* the causal agent of cucumber anthracnose, and reduced the size of stem lesions caused by *Fusarium oxysporum* f. sp. *cucumerinum. Pseudomonas* spp. also have been used to induce resistance to some root-rotting organisms, such as *Pythium* spp. Plant growth-promoting rhizobacteria induce resistance to *Colletotrichum orbiculare* and cucumber virus (Baker and Paulitz 1996).

Some of the benefits of intensified use of cover crops and reduced tillage in controlling soilborne diseases may be due to a buildup of sufficient antagonistic organisms to the disease that it is not as potent. As pointed out earlier, we can expect considerable increase in soil microorganisms as organic matter is added, and antagonistic organisms can be expected to increase as the general bacterial population increases. It is interesting to note that usually "healthy" soils are obtained as lime is applied to raise the soil pH in a range of 6.0-7.0. Some of this effect may be due to providing satisfactory growing conditions at this pH (reduced Al, increased Ca, improved porosity), but it appears some of this response is due to increased bacterial activity at these pH values.

Cases have been reported where little or no disease is apparent despite the presence of ample pathogen and plants susceptible to it. Strangely enough, some of these so-called suppressive soils have developed under monoculture. Although monoculture favors the development of certain diseases, continuous use over long periods appears to encourage organisms that are antagonistic to the disease organism.

Although some suppressive soils may come about through monoculture, continuous growing of a single crop is not the best way to reach this goal, since similar suppression can be obtained by using organic matter. Antagonistic microorganisms have been isolated from suppressive soils, and increasing the organic matter evidently is a viable means of increasing these organisms in soil.

The supposition that suppression is largely due to antagonistic microorganisms is supported by experiments that indicate loss of suppression due to treatments that kill most of the organisms but restoration by reintroducing specific organisms. Much of the work has been done with the suppression of *Fusarium oxysporum*, a fungus that can cause serious diseases in many plant species in practically all botanical families except the Graminaceae. The work, which covers about 40 years, has shown that many different types of organisms can help control fusarium diseases, with *Pseudomonas fluorescens*, *P. putida*, and nonpathogenic strains of *F. oxysporum* some of the leading contenders. The ability of nonpathogenic strains of *Fusarium* spp. to control the disease has been attributed to (1) competition for nutrients, (2) competition for infection sites, and (3) induced resistance. The more recent work indicates that a combination of selected strains of *F. oxysporum* plus the fluorescent pseudomonads have a good chance of providing suppressiveness for other soils or media.

The outlook for utilizing suppressive soils appears most promising for pot cultures, where a limited amount of media is needed and crops have high value (Alabouvette et al. 1996).

Insects

A large group of insects live in the soil, although some of these only for short periods, usually during the winter months. Fortunately, most of them are beneficial in that they tend to shred OM into smaller fragments, making them more readily decomposed by other soil inhabitants. But some, especially in the larval stage, feed on plant roots and can cause serious damage. The damage is often accentuated because the injury induced by feeding serves as an entry for several diseases. Other soil-inhabiting insects that can be damaging are springtails (*Collembola* spp.), beetle larvae (*Coleoptera* spp.), fly larvae (*Diptera* spp.), various cutworms (*Lepidoptera* spp.), wireworms (*Elateridae* spp.), ants and termites (*Hymenoptera* spp.), white grubs, larvae of *Scarabaeidae Coleoptera* spp., lesser cornstalk borers [*Elasmopalpus lignosellus* (Zeller)], seed corn maggot [*Hylemya* (Delia) *platura*], the larva of *Bothynus, Strategus, Cyclocephala,* and *Phyllophaga* spp. listed as white grubs, and the scale insects ground pearls (*Margarodes* spp.). Although a healthy balance of soil organisms can be useful in limiting damage to seeds and seedlings, these benefits appear to be minor in comparison to those of cover crops as they support beneficial insects capable of keeping plant insects in check.

The beneficial insects, consisting of parasites and predators of harmful insects, are nurtured by the cover crop and help to control the harmful ones by reducing their numbers or effectiveness. Intensive growing of cover crops, which favors the beneficials, when combined with reduced tillage has increased OM and SOM. The extra organic matter has simplified pest control. In case of insects, the system has lessened the damage not only from soil-inhabiting insects but also from those that live out their life above ground. Programs of intensified use of cover crops have paid off with lessened need for insecticides by California growers of almonds and walnuts, cotton growers in Georgia, Mississippi, and South Carolina, pecan growers in California, and peanut, cotton, and vegetable farmers in Georgia. Specific cover crops that help support beneficial insects are crimson clover, cahaba white vetch, hairy vetch, woolypod vetch, rye, sorghum, buckwheat, and sweetclover (Phatak 1998).

Weeds

Soil organic matter can also help to control weeds. Control can result from an indirect effect whereby the planted crop is able to get off to a fast start due to the many benefits of organic matter. The fast start enables the crop to restrict weed growth by shading, smothering, or overcrowding it. Without ample organic matter, the weeds would probably have the upper hand because, as natural survivors, they probably are better able to cope with adverse conditions of moisture, temperature, acidity, and nutrient levels.

Improved weed suppression is obtained if proper cultural practices are combined with the benefits of organic matter. Every effort should be made to restrict the addition of weed seeds by using plant seeds free of noxious weeds, preventing weeds from going to seed by timely mowing or use of herbicides, and killing of weed seeds in manure by composting.

Cover crops, which have the advantage of adding to the OM, can be effective in controlling weeds by shading, overcrowding, and competing with them. Some cover crops also control weeds because they are a source of allelopathic compounds that interfere with weed germination or their later growth. Some specific cover crops that are beneficial for different regions in the United States are given in Table 5.1.

TABLE 5.1. Cover crops suitable for suppressing weeds in different regions of the United States

Region	Cover crops
Northeast	sorghum-sudangrass*, ryegrass**, rye, buckwheat
Mid-Atlantic	rye, ryegrass, oats, buckwheat
Mid-South	buckwheat, ryegrass**, oats, subterranean clover
Southeast Uplands	buckwheat, ryegrass**, subterranean clover, rye
Southeast Lowlands	Berseem clover, rye, wheat, cowpeas, oats
Great Lakes	Berseem clover, ryegrass**, rye, oats
Midwest Corn Belt	Rye, ryegrass, wheat, oats
Northern Plains	Medic, rye, barley
Southern Plains	Rye, barley
Inland Northwest	Rye, wheat, barley
Northwest Maritime	Ryegrass, lana woolypod vetch, oats, white clover
Coastal California	Rye, ryegrass, berseem clover, white clover
California Central Valley	Ryegrass, white clover, rye, lana woolypod vetch
Southwest	Medic, barley

Source: Based on data in Chart 1, Top regional cover crop species, in A. Clark (Coordinator). 1998. *Managing Cover Crops Profitably,* Second Edition. Sustainable Agriculture Network Handbook Series Book 3. Beltsville, MD: Sustainable Agriculture Network, National Agricultural Library.
*Sorghum-sudangrass hybrid
**Annual ryegrass

Nematodes

Of the many species of nematodes, only about 25 are serious pests of plants. Some of the more important nematode pests are dagger (*Xiphinema* spp.), root-knot (*Meloidogyne* spp.), root lesion (*Pratylenchus* spp.), stubby root (*Trichodorus* spp.), sting (*Belonolaimus* spp.), stunt (*Tylenchorhynchus* spp.), lance (*Hoplolaimus* spp.), cyst (*Heterodera* spp., *Globodera* spp., *Punctodera* spp.), potato cyst (*Globodera pallida* and *G. rostochiensis*), soybean cyst *(Heterodera glycines),* burrowing *(Radopholus similis),* awl (*Dolichodorus* spp.), and bulb and stem (*Ditylenchus* spp.).

Some nematodes cause considerable damage because their infestation allows various diseases to proliferate. Blackshank, a disease of cotton, caused by *Phytophthora parasitica* var. *nicotinae,* develops only if the cotton is infested with cotton nematode *(Meloidogyne incognita acrita)* as well as the disease organism. Plants not infested with the nematode did not develop the disease even when inoculated with the fungus. A number of viral diseases (tomato black ring, raspberry ring spot, cherry leaf roll, strawberry latent ring spot, arabis mosaic, grapevine fanleaf, pea early browning, tobacco rattle virus, and brome mosaic are transmitted by nematodes (Bruehl 1987).

The millions of nematodes found per square meter consist largely of the nonparasitic types that may live on organic matter, soil microorganisms, or soil fauna. These types that feed on several forms of fauna, including other nematodes, are largely responsible for keeping the parasitic species in check. Amoebas and bacteria also feed on nematodes.

The use of organic matter to control damaging nematodes is being examined more closely as methyl bromide, the primary fumigant for controlling damaging nematodes, is being phased out. Increased cover crop use combined with reduced tillage has been able to keep moderate or low infestations of parasitic nematodes from causing serious damage (Phatak 1998). It has been well established that interplanting with marigolds or canola helps keep a number of nematodes in check. The two crops have been mainly used in gardens or noncommercial operations, but the increased popularity of canola oil might make canola a suitable plant for strip cropping in commercial operations in certain climates.

Chapter 6

Adding Organic Matter

The many benefits of SOM make it mandatory to increase the amounts of OM added to the soil and to maximize the effectiveness of the additions in every possible way. Due to the continuous decomposition of organic matter that is hastened as soils are cultivated, regular additions are needed to maintain adequate levels.

The benefits of SOM can be augmented by (1) increasing the amounts of organic materials returned to the soil, (2) placing them so that maximum benefits are obtained, and (3) optimizing the conditions favoring the conversion of organic materials to SOM.

There are essentially two different ways in which organic matter can be introduced into the system: (1) it can be raised in place in the forms of cover crops, sods, pastures, and hays or by utilizing residues of crops raised for other purposes, or (2) it can be brought in as manure, composts, organic fertilizers, peats, various crop residues, biosolids, or organic wastes. Growing it in place is by far the more economical method, but types of organic matter suitable for maintenance are more limited.

This chapter outlines the benefits and process of growing organic matter in place. Organic matter not grown in place is covered in Chapter 7. The placement of OM is considered in Chapter 8, and optimizing conditions favoring the conversion to SOM in Chapters 9 and 10.

GROWING ORGANIC MATTER IN PLACE

For many growers, adding extra organic matter will be economically possible only if they grow it in place, avoiding costs of hauling and handling materials. Many different kinds of plants can be grown economically which provide materials that benefit soils and crops. Ideally, the amounts produced will increase SOM, but even minimal amounts that do not increase SOM tend to provide benefits by reducing erosion and improving certain soil properties such as water infiltration and storage as well as providing sufficient gaseous exchange in the soil.

Hays, Pastures, and Forages

One of the best ways of increasing SOM is by growing long-term sod crops such as pastures, hays, and some forages. Many of these plants have heavy root growth that extends deeply into the soil. Grasses can add a ton of dry matter per acre per year from roots alone, and many grassland soils contain over 5 tons per acre of roots while having only 1-2 tons of stems and leaves. Channels opened by roots tend to move moisture and air deep into the soil. The movement and storage of water deep in the soil accounts for much greater water retention under sod as compared to tilled crops, and is probably responsible for a good part of the increased yield of cultivated crops following sods compared to that of continuously cultivated crops.

Various plants have been used for such purposes. Alfalfa, red clover, ladino clover, bird's-foot trefoil, Kentucky bluegrass, coastal bermudagrass, ryegrass, timothy, and meadow fescue are used extensively in temperate zones as hayfields and pastures, while grasses such as bahia, guinea, napier, para, and pangola are common in tropical or subtropical areas. The temperate-zone grasses are often mixed with white clover.

Some forage crops can provide similar advantages. Sorghum-sudangrass hybrids, berseem, crimson, subterranean, red, white, and sweet clovers and woolypod vetch provide good forage in temperate zones, although the sorghum-sudangrass hybrids should not be used as forage when young (up to 24 inches tall), drought stressed, or killed by frost as they can induce prussic acid poisoning in livestock. All of these forages can be used as annuals, but the subterranean clover can reseed itself, red clover can be a short perennial or a biennial, white clover can be a long-lived perennial, and the sweet clovers can exist as biennials. Forages lasting two or more years can supply substantial amounts of OM, providing some of the benefits of the pasture and hayfield sods. All except sorghum-sudangrass hybrids are legumes that can fix large quantities of N.

Rotations

Using hay, pasture, and forage crops for OM additions can be a viable choice for growers of livestock but is not practical for most growers. Growing these crops in rotation with cultivated crops may be a practical solution for some growers, especially if they have some livestock. However, rotations of primary crops or primary crops with cover crops may be viable options for many other growers, especially since rotations supply other advantages such as the suppression of several pests.

Rotations and SOM

Continuous cropping results in depleted SOM, but the amounts remaining vary with different crops. As was seen in Table 2.3, the SOM loss depends on the types of crops used in the rotation. Rotation of crops tends to increase organic matter, especially if a close-rowed crop, such as wheat, is included with a wide-row crop, such as corn. The increase in SOM is probably due to the reduction in tillage for close-rowed crops, which limits the amount of O_2 available for decomposition of the SOM.

The value of rotations for increasing SOM has been established from years of experience and some long-term experiments. Typical results from the long-term Morrow or Breton experimental plots follow.

The Morrow plots, established in 1876 in Illinois, showed that rotations induced marked changes in SOM over time. As shown in Table 6.1, SOM dropped most markedly if corn was grown continuously in the 77-year period from 1876 to 1953, but the loss was substantially reduced if a rotation of corn: oats was used, and the loss was least if the rotation was changed to corn: oats: three years pasture. The loss of SOM was eliminated if farmyard manure, lime, and phosphate were added to the corn: oats: three years pasture rotation. Red clover was used as a pasture for the last 52 years reported (The Morrow Plots 1960).

The more recently published report of the Breton long-term experiments, conducted over a 51-year period (1939-1990) at the Breton Classical Plots in Breton, Canada, indicate that average aboveground

TABLE 6.1. Changes in SOM of the Morrow plots as influenced by rotation and applications of manure, lime, and phosphate

	Treatment			
	1904		1953	
	None	M+L+P*	None	M+L+P*
Rotation	(%)	(%)	(%)	(%)
Continuous corn	4.0	3.8	2.4	3.1
Corn→oats	4.3	4.4	3.3	4.4
Corn→oats→3-year pasture	5.0	5.1	4.0	4.9

Source: *The Morrow Plots*. 1960. Illinois Agricultural Experiment Station Circular #777, University of Illinois, Urbana, IL.
*Manure (M) was applied before corn, at a rate equal to the dry weight of crops removed. Lime (L) was applied five times since 1904. A total of 6.6 tons of rock phosphate (P) was applied from 1876 to 1925 and none thereafter.

C production of 514 lb/acre obtained in a wheat: fallow rotation without fertilizer or the 963 lb obtained in the rotation with fertilizer was insufficient to maintain soil organic carbon (SOC). However, the 1,078 lb produced in the same rotation with manure increased SOC. More aboveground C was produced in a rotation of wheat: oats: barley: hay (mainly alfalfa). This rotation even averaged 762 lb SOC per acre without fertilizer, 1,635 lb with fertilizer, and 1,531 lb with manure (Izaurralde et al. 2001).

It is interesting to note that the long rotation increased SOC even without fertilizer or manure. Evidently, reduced tillage associated with a sod crop and small grains in a cool climate slowed SOM decomposition sufficiently that relatively low inputs of OM were sufficient to boost SOM (SOC = 1/2 SOM).

Rotations can be an effective means of increasing yields, even though SOM may not be fully maintained. Part of the yield increases obtained with rotations is due to improved nutrient availability and suppression of pests, but many other factors, such as decreased erosion, improved soil moisture, and better storage and movement of water and air resulting from improved soil structure appear to be involved.

Rotations and Nutrient Availability

Availability of nutrients is increased in some rotations because deep-rooted legumes and nonlegumes in the rotation may bring nutrients from deeper portions of the profile and make them available for shallow-rooted primary crops. The larger amounts of OM and SOM usually associated with rotations tend to provide for more efficient recycling of nutrients, as there is a more orderly release of nutrients for the primary crop. Nitrogen efficiency is probably increased more than that of any other element.

Nitrogen utilization in rotations. Better N utilization (NUE) is cited as an important advantage of rotations. Better NUE appears to be due to a combination of (1) reduction of N losses by erosion, and (2) more efficient N fixation by legumes because the N is added to nonlegumes in the rotation. The increase in N fixation may be due to better survival of the symbiotic organisms fixing N in a crop rotation than in monoculture as well as the tendency of the organism to use available N rather than fixing it if N is applied directly for the legume. If an off-season crop, such as a cover crop, is included in the rotation, NUE is further increased, since a good part of the N left over from the primary crop is recovered.

Rotations and Pest Control

The beneficial effects of reduced disease, insects, and nematodes have been associated with long rotations of primary crops, but short-term rotations also are beneficial. In Illinois, the popular corn: soybean rotation yields about 10 percent more corn and soybeans than if either crop is grown continuously (Nafziger 1994). Some of the increase in yield in short-term rotations may still be due to better pest control, but other factors, such as the use of primary crops that provide more N or residues, or require less cultivation, appear to be helpful. But rotations that include cover crops and/or sods add more organic matter to the system by reducing the extent of cultivation and adding OM.

Crop rotations tend to reduce weed problems, and rotations can be selected to help control weeds. Rotations that include alfalfa, annual ryegrass, barley, berseem clover, buckwheat, cowpeas, oats, winter rye, and winter wheat tend to limit problems of weed control and should be used wherever possible.

The OM advantages gained from long-term sod forage and hay crops can be utilized by several years of growing row crops, after which the organic matter can be replenished by growing sod, forage, and hay crops for a few years. Such arrangements were rather common on family farms up until about 50 years ago, but the short-term economics associated with monoculture eliminated such programs on vast acreage. The shift to monoculture was hastened by the elimination of farm animals as power sources, freeing large amounts of land needed to grow forage for the animals. Today, it is impractical for many growers to use these crops exclusively as a means of optimizing SOM, but some of them can use these or other beneficial crops in a rotation with cash crops.

Rotation can effectively reduce pest problems of many crops. By introducing nonsusceptible or more resistant crops, the numbers of pests are reduced while other organisms are favored. These additional organisms tend to reduce the numbers of the pest through consumption and inhibition as they compete with the harmful organisms for nutrient elements and energy.

Rotations are especially beneficial in restricting plant diseases when they help increase the amount of organic matter in the soil. Not only are greater numbers and more diverse populations encouraged, but the organic matter tends to alter the soil so that the crop is favored. This is most strikingly demonstrated with facultative disease organisms capable of growing readily in both aerobic and anaerobic atmospheres. These organisms, such as *Pythium* spp. or *Fusarium* spp., can grow readily in soils that are overly wet, but most crop plants do poorly under such conditions. The additional organic matter, by improving porosity, favors the aerobic plant, allowing it to grow normally and preventing the disease organism from gaining the upper hand.

Crop rotation has been used for many years to control nematodes. To be effective, the rotation needs to include crops not susceptible to the damaging nematode for a long enough period to reduce the numbers to noneconomic levels. For example, cotton can be grown successfully for two to three years in a row in the presence of the root knot nematode *Heterodera marioni* (Cornu) Goodey if nonsusceptible crops (sorghum, small grains, alfalfa) are grown for a period of three years. On the other hand, six years of nonsusceptible crops (alfalfa, sweet clover, beans, potatoes, cereals, and vegetable crops) have to be grown between a single crop of sugar beets to avoid economic damage from the sugar beet nematode (*Heterodera schactii* Schmidt) (Leighty 1938).

Rotation of crops has helped cotton and peanut farmers, in Georgia enabling them to greatly reduce their use of pesticides. The rotation substitutes a few other cash crops for some of the cotton and peanuts. Including a strong cover crop program and using minimum tillage has helped maximize the effectiveness of the rotation. Two to three years in a program of intensified use of cover crops plus reduced tillage has about eliminated damping off of cotton, peanuts, and vegetables due to attacks of *Rhizoctonia solani, Pythium myriotylum, P. phanidermatum, P. irrglare,* and *Sclerotium rolfsii* (Phatak 1998).

Rotation of vegetable crops with crimson clover and winter rye has enabled Georgia growers of cabbage, cantaloupes, cucumbers, eggplants, lima beans, snap beans, and southern peas to greatly reduce fungicide use. Planting through residues of primary and cover crops left as mulch or strip tilling of the rye and crimson clover maximized the value of added OM (Phatak 1998).

Types of Rotations

Various types of rotations are being used. A rotation of rye or tall fescue used as winter forage with corn during the summer effectively reduces erosion and provides for good yields of corn, providing all the rye or fescue is killed prior to planting the corn. A rotation of three years alfalfa: two years vegetables has been practical in several areas. A number of crop rotations are being used extensively: corn: sorghum in the midwest; small grains: canola in the Canadian prairie provinces; wheat: soybeans in the East and South; wheat: sorghum in the plains. A three-crop rotation of corn: wheat: sorghum is popular in North Carolina. Rotations in the Corn Belt may include corn: wheat: clover; corn: oats: clover; or two years corn: oats: clover. Two years of pangola grass in a rotation with two years of stake tomatoes has greatly increased production of tomatoes in Florida. Longtime pastures of bahia grass or native grasses also have been used in Florida for one or two years of tomatoes or watermelons. Rice in rotation with sugarcane improves sugarcane production in Florida.

Rotating Crops Using Strip Cropping

Although most rotations utilize the entire area for either the primary or organic matter crop, large areas are devoted to strip cropping. The strips may vary in width from a few feet that are primarily grown for wind erosion control to about 200 ft. Minimum practical widths for most crops are about 50 ft. A cultivated crop is usually alternated with crops that need little or no cultivation (sods, hayfields, small grains). The small grains allow annual crop changes in the strip, while a rotation with sods and hayfields may require changes every two or more years.

Field strip cropping is laid out in approximately parallel strips usually perpendicular to the general slope. Unless applied to regular slopes, they are much less effective in reducing erosion than strips that generally follow the contour of the land. The latter method is often combined with various mechanical methods of erosion control, such as terraces and diversion ditches. Both methods of strip cropping offer advantages in conservation of water and maintenance of SOM when cultivated crops are rotated with sod or forage crops, hay, or small grains, but contour strip cropping is usually superior.

It is not always possible to rotate crops in strips used for erosion control. Because of slope or the erodibility of the soil, some conservation strips may have to remain in a permanent sod. But where erosion is not that serious, it usually is desirable to rotate the crops in contour strips to gain better pest control and to benefit from improved SOM contents.

Rotations with Cover Crops

The option of including pastures, sods, hays, or forages may not be suitable for most growers, but there are very few growers who would not profit from the introduction of off-season cover crops in the rotation. Green manure or cover crops are OM sources suitable for short growing periods. Off-season cover crops utilize the land at times when no primary crop is grown. They can be helpful in reducing wind and water erosion, especially during the winter period when the soil may be bare. They greatly improve soil aggregation and may add considerable OM. Their use helps save nutrients that may have been left over from the primary crop (scavenging).

The value of cover crops is influenced by the kind of crop, its age when it is plowed under or terminated, amounts of nutrients left over from the previous crop, soil type on which it is grown, climate, and in the case of the many legumes, the soil's pH and its content of Ca, P, K, Mg, and B. The value of cover crops for increasing SOM is subject to appreciable variation. Often, these crops are in the ground for such a short time that there is little or no increase in SOM content. Even though there is no increase in SOM, they can have a positive effect by reducing erosion and helping to hold existing SOM levels or by reducing the rate of loss.

The effect on SOM improves if the crop is allowed to approach maturity or at least reach the blossom stage before it is terminated, as older crops produce more OM that is slower to decompose. The nonlegumes have an advantage over the legumes by resisting decomposition, but the lack of N may need to be compensated by additions of manure or fertilizers.

Catch or cover crops cannot be expected to benefit structure as well as the longer pasture, hay, or sod crops, but temporary effects, especially in cool climates, can be helpful. Evidently, the organic matter provides enough cementing material to produce some extra aggregates of the silt and clay, providing extra air spaces—at least for short periods. The benefits of short-term crops may be partly due to their preservation of existing structure by lessening the harmful effects of heavy rains or strong winds.

Growing the Cover Crop for More Biomass

The yields of cover crops can be greatly increased by letting them grow for longer periods. The potential for obtaining full yields of fall cover crops is enhanced by seeding them in the primary crop after most of the growth of the primary has been made. Usually, maximum dry matter of the cover crop will be produced by the time it is in full bloom, and there is little or no benefit to leaving it longer unless viable seed is needed to reseed the cover crop. Terminating the cover crop before bloom, however, can reduce the OM produced, and should be avoided wherever possible. But there are three drawbacks in growing the cover crop to the bloom stage or beyond. (1) The C:N ratio narrows as the crop matures, making it necessary to have more available N present when planting a primary crop following a near-maturity cover crop as compared to following a juvenile cover crop. (2) The cover crop raised to bloom stage will remove much more moisture than one terminated several weeks earlier, making it difficult during dry seasons to successfully start the primary crop. (3) There may not be enough time between terminating the cover crop and planting the primary crop to dissipate allelochemicals present in some cover crops.

These drawbacks can seriously affect the primary crop, but may not be present in all cases or may be corrected by proper measures. The first problem, N shortage, can be greatly reduced by adding enough quickly available N to narrow the ratio. Soil tests for available N can help decide if the application is needed and when to apply it. The second problem, reduced moisture, occurs only during a dry spring, when allowing a cover crop to bloom before terminating it can delay the planting or the emergence of the primary crop. The problem of moisture shortage, if it does exist, is easily handled if irrigation is available. Drip irrigation allows a simple solution of the two problems of moisture and nutrients by adding the quickly available N through the irrigation with little waste of water. The third problem also may not exist except with certain cover crops or residues. The presence of allelochemicals also may not be a problem because most of the organic matter is left on the surface and is not in intimate contact with the developing plant. The possibility of intimate contact can also be reduced by mechanically moving a small strip of the organic materials just ahead of the planter away from the planting area.

Cover Crops and Nutrient Availability

Cover crops can increase the amount of nutrients available for the primary crop. Leguminous cover crops fix appreciable quantities of N for the succeeding crop. Both legumes and nonlegumes tend to benefit the soil nutrient status by maintaining nutrients in a slow-release state so that they are not readily lost to leaching or denitrification. The improved aggregation resulting from the addition of OM helps develop a crumb structure that increases air and water storage as well as improving water infiltration and drainage. The favorable air and moisture resulting from crumb structure favors both symbiotic and nonsymbiotic N fixation.

Legume cover crops can be a great source of N because they can fix appreciable quantities (75-250 lb/acre) of N from the air. Some cover crops, such as berseem clover, hairy vetch, subterranean clover, and woolypod vetch can fix 200 or more pounds of N per acre, but the amounts are often less—usually closer to 100 lb of N per acre. The mineral elements Ca, K, Mg, B, and Mo are closely involved in symbiotic N fixation. The Ca, K, and Mg are held in an available form in the exchange complex and, as has been pointed out, SOM can be an important part of CEC. A large CEC can help ensure sufficient supplies of these essential cations.

Calcium appears to be the most important cation necessary for symbiotic N fixation, at least for some legumes. In some cases, such as alfalfa, sweet clover, and red clover, a near-neutral pH is also necessary. Subterranean or white clover and many tropical legumes can fix N at much lower pH values, even down to a pH of about 4.0 if there is sufficient Ca in the soil. Actually, very small amounts of Ca are needed for some of these that are tolerant of low pH. It is not certain whether the higher pH values required by some temperate-zone legumes are due to their inability to extract sufficient Ca unless amounts are high, or whether Al or Mn, normally increased as pH is lowered, may be elevated enough to depress growth of these plants, which are sensitive to high levels of these elements.

While a good supply of most nutrient elements tends to increase N fixation, that of N usually decreases it. An ample supply of available N (NH_4-N and/or NO_3-N), whether from fertilizer, manure, sludge, compost, or large amounts of decomposing OM, will reduce the amounts of N fixed by the symbiotic process as the host plant utilizes the readily available forms.

The very large amounts of N fixed by many legumes are obtained in soils of good tilth with ample P, K, and Ca. If the crop has not been grown before, it has to be inoculated with a strain of N-fixing organisms that is compatible with the legume.

The amount of N fixed by the legume and the availability of the fixed N depends on the availability of soil N, Ca, Mg, P, and K, the type of legume, and its stage of growth when terminated. Annual legumes, such as cowpea, crimson clover, hairy vetch, field pea, and subterranean clover, will break down and release N quicker than the perennials. Young plants prior to bloom will tend to have 3.5-4 percent N in the aboveground dried material, 3-3.5 percent at time of flowering, and appreciably less in older plants if the seed crop is removed. The younger plants, which have about 4 percent N, tend to decompose quicker than mature plants, while older plants with seeds will release N at the slowest rate. Although N concentrations are highest in young plants, total amounts fixed by the crop when young tend to be low due to lower biomass at early stages. Terminating the legume at bloom tends to provide the largest amount of N that is released in a reasonable period.

Nonleguminous cover crops, although they do not fix N, can provide substantial amounts obtained as they scavenge leftover N or in rare cases from fertilizer applied for the cover crop. The amount of N also varies with the crop, the amount of available N, and the stage when it is terminated. As with legumes, the highest concentration of N will usually be present when the crop is young, although the N concentrations are usually appreciably lower than for legumes (2-3 percent before flowering and 1.5-2.5 percent after flowering for grasses compared to 3.5-4 percent for young legumes and 3-3.5 percent at flowering for legumes).

Because certain nonlegumes can produce greater biomass than some legumes, the total N produced in a given area may at times be greater with the nonlegume. An example of greater N production by sorghum, a nonlegume, as compared to velvet beans, a legume, grown under tropical conditions during the summer months in Guatemala is presented in Table 6.2.

Among their other attributes, cover crops serve a useful purpose in mopping up leftover nutrients from the previous crop. Without their use, most of these nutrients may be lost to leaching or erosion and in many situations increase pollution problems. Fortunately, in some cases the nutrients scavenged by the cover crop are sufficient to pro-

TABLE 6.2. Comparative production of dry matter and nitrogen by sorghum and velvet beans grown in Guatemala during the summer months*

Cover Crop	Dry weight (lb/acre)	Nitrogen	
		%	lb/acre
Sorghum	10,305	1.25	143
Velvet beans	2,988	4.20	116

Source: Private communication from Roberto Dubon Obregon, Central American Produce, Zacapa, Guatemala.
*Results from an experiment conducted on the Central American Produce farm in Zacapa, Guatemala.

vide ample growth. But often, particularly on poor soils, nutrients left from the previous crop are insufficient to grow a satisfactory cover crop. The problem is aggravated with nonlegumes, since lack of N is usually responsible for poor performance of some cover crops.

The nonlegumes may not contribute any N to the fertility program unless some is salvaged from amounts left over from the previous crop or is absorbed as the previous crop decomposes. Planting legumes with nonlegumes can solve the problem of poor N contributions by the nonlegume. The combination may have advantages over some legume cover crops, as the legume supplies N that increases the growth of the nonlegume, often supplying more OM than is obtained from the legume by itself. If the nonlegume is a grass, there may be added benefits from longer-lasting improvements in soil structure due to the large fibrous root systems of many grasses.

It may not be possible to plant legumes with late-planted nonlegumes in some areas because there is not enough time to establish the legume before cold weather sets in. In such cases, cool-weather nonleguminous cover crops, such as rye or ryegrass, can produce more OM if the cover crop is fertilized with N. On the coastal sandy soils of southern New Jersey, greater response of the cover crop was obtained if K was included with N applications.

In several studies, the senior author found that some of the fertilizer (10-10-10) intended for the succeeding pea crop could profitably be applied to the late-planted cover crops (rye or ryegrass). The production of OM in most cases more than doubled, while the yield of the primary crop that followed (English peas) increased slightly. Yields of English peas may have been increased by the additional organic materials produced, but the conversion of inorganic N in the 10-10-10 to organic forms that withstand leaching by spring rains on these light soils may also have played a role.

As has been pointed out earlier, the N contents of organic materials, while useful for estimating N needs of the subsequent crop, are not reliable indices as to whether extra N will be needed, or, if needed, when it should be applied. The release of available N from cover crops or any other organic material depends not only on the type of material and its age when terminated, but also on soil nutrient availability, soil pH, soil porosity, soil biota, and climatic conditions (temperature and moisture).

Trials have indicated considerable N contributions to primary crops by legume and nonlegume cover crops, but increased yield of the primary crop from added fertilizer N may still occur, evidently because N release from the cover crop is insufficient or does not coincide with the crop's need. Although the need for extra N is much less common if a legume cover crop is used, adding N after incorporating leguminous cover crops often will increase yield of the primary crop. As has been pointed out, soil tests for available N (NO_3-N + NH_4-N) can be helpful in deciding whether additional N is needed.

The senior author's experience using repeated soil tests for various crops indicates that response to added N despite incorporation of a legume cover crop is rather common and usually is due to a shortage of available N for a relatively short period. In the Northeastern states, this period usually lasts only for a few weeks during the early spring, shortly after planting the primary crop.

In a trial of several cover crops (rye, oilseed radish, oat, and red clover) for corn in south central Ontario, all cover crops were effective in sequestering available soil NO_3 present after wheat harvest, but only red clover reduced the N requirements of corn. The red clover produced the largest biomass and total amount of N. Despite the addition of about 54 lb N per acre for each of the three nonlegumes in a three-year period, they did not increase N availability. The response to a side dressing of 134 lb N per acre was greater with rye or oats than with oilseed radish and least with red clover cover crops. It is interesting to note that despite an addition of over 175 lb N per acre in the three-year period by red clover, there was still response to the side-dressing of N. It was concluded that a pre–side-dress test for soil NO_3-N was more accurate in predicting response to an N side-dressing than a test made at planting. Evidently, allowing time for some of the OM to decompose gave a better idea as to amounts of N available for the crop (Vyn et al. 2000).

Cover Crops As Aids for Pest Control

Including cover crops in the rotation can be helpful in reducing pest problems. A good part of the disease and insect control obtained in Georgia with the extended use of cover crops appears to derive from the additional OM supplied by the cover crops, aided by reducing the extent of tillage. The combination apparently has led to an increase in the SOM, but this has taken several years to accomplish (Phatak 1998).

Evidently, some reduction of disease can take place with the use of cover crops before SOM is substantially increased, since reduction in verticillium wilt of potatoes in Idaho has been obtained with only one year's use of sudangrass as a cover crop. But generally, substantial reduction in diseases with increased cover crop use requires more time. The two- to three-year control of damping-off listed for Georgia is common for moderate infections, but longer periods may be needed for heavy infestations. It took about six years to control serious stem lesion losses of potatoes in Maine induced by *Rhizoctonia solani*. The additional time may be necessary to build up enough SOM to alter several soil qualities that affect soilborne diseases (Clark 1998).

Insect control is aided by use of cover crops that support substantial populations of beneficial insects and mites. There has been a trend in recent years to limit the use of insecticides and to depend more on natural enemies of pest insects. The trend has been motivated by reasons of pollution control, reducing costs, and limiting introduction of chemicals potentially capable of harming humankind, but it has been made possible by harnessing the activities of beneficial insect and mites that feed on or disrupt the lives of the harmful insects.

To obtain the full benefits of beneficial insects, it is important that a protective habitat be maintained for them at all times during the growing season so that sufficient numbers can be ready to control an invasion of harmful insects. Cover crops can be an important link in this survival chain because they can harbor a number of beneficials throughout the year. There are a number of generalist predators, such as insidious flower bugs *(Orius insidiosus)*, bigeyed bugs (*Geocoris* spp.), and lady beetles *(Coleoptera coccinellidae)*, which can increase in great numbers on the nectar, pollen, thrips, and aphids of vetches, clovers, and some cruciferous cover crops (Phatak 1998). Leaving enough cover crops by use of strip tillage, as windbreaks, or on edges of fields when the cover crop is usually terminated will help preserve a nucleus of beneficials that can quickly generate sufficient numbers to control the invading insect pests.

Cover crops also play an important part in weed control. Dense plantings tend to exclude light, which delays or inhibits growth of weeds. Rye, barley, wheat, buckwheat, and sorghum-sudangrass hybrids are especially suited for this purpose because they are quickly established and tend to take up moisture and nutrients needed by weeds, smothering the weeds.

Once terminated, residues of the cover crop, left as mulch, tend to suppress the emergence of weeds. Some cover crops, such as rye, have the added advantage of allelochemicals that are capable of inhibiting the growth of many annual dicotyledonous weeds. Growth restriction of monocotyledonous (grasses) weeds by allelochemicals in the cover crop residue is much less reliable. The growth of grasses and any late growth of the dicotyledonous weeds often can be kept in check by a late application of metribuzin (Putnam 1990).

Examples of Rotations with Cover Crops

Many different rotations with cover crops are being used successfully. A typical example is a wheat: rice rotation used in India. In a subtropical, semiarid climate in the Punjab province of India, a four-year rotation uses irrigated rice in the first year followed by three years of wheat: sesbania: cowpeas. The rotation with the cover crops of cowpea and sesbania provided advantages over the rotation of wheat: irrigated rice in allowing a 50 percent reduction of N applications for rice and a 25 percent reduction for wheat without sacrificing yield.

A number of rotations with cover crops used in the United States are presented here. The rotations vary with different crops and locations. In a corn: soybean rotation, rye can follow corn, and hairy vetch can be planted after soybeans in Zone 7 (minimum temperatures 0 to 10°F [-18 to -12°C]). In cooler climates, the rye needs to be seeded as soon as the corn is harvested or overseeded in early summer at last cultivation of corn or early leaf fall (yellow leaf stage) of soybeans. In a corn: rye: soybean rotation, a year of small grains is added. A short-season bean crop is planted after the soybeans, but if time does not permit, a small grain such as rye or barley may be planted. An alternative rotation of corn: crimson clover (allowed to go to seed): soybeans: crimson clover (reseeded): corn is suggested for the lower mid-South. The crimson clover is allowed to go to seed before planting beans, but needs to be killed before planting corn the next spring. In the upper Midwest, a three-year rotation of corn: soybeans: red clover has been used successfully. Red clover or sweet clover can be frost seeded in the wheat in mid-March and allowed to grow until dormant in late fall, to be followed with corn in the spring. A legume cover crop such as hairy vetch, planted immediately after harvesting the small grain can be substituted for red or sweet clover.

If moisture is ample, cover crops of annual ryegrass, crimson clover, or combinations of rye and crimson clover can be overseeded in the fall vegetable crop and provide needed cover during the winter. A three-year rotation of winter wheat: legume: interseeded legume: potatoes works well for eastern Idaho potato production, and a one-year rotation of lettuce: buckwheat: buckwheat: broccoli: white clover or annual ryegrass is very effective in the Northeast. In California, a four-year rotation of vetch: corn: oats: dry beans: common vetch: tomatoes: sorghum-sudangrass hybrids: cowpea: safflower is widely used.

In dryland areas, both a seven-to-thirteen-year rotation of flax: winter: spring barley: buckwheat: spring wheat: winter wheat: alfalfa (up to six years): fallow, and a nine-year rotation of winter wheat: spring wheat: grain/legume interseed: legume green manure/fallow: winter wheat: spring wheat: grain legume interseed: legume: legume are popular.

Both one-year and multiyear rotations are used for cotton. In the one-year cotton rotation, rye is combined with crimson clover, hairy vetch, calhaba vetch, or Austrian winter peas and planted in mid or early October to allow the legume to become well established before the cooler winter temperatures restrict growth. The cover crop is killed in late April, and no-till cotton is planted within three to five days, providing moisture is satisfactory. Subterranean clover, Southern spotted burclover, Paradana balansa clover, and crimson clover are used as the legumes in the multiyear rotation of reseeding legume: no-till cotton: legume: no-till cotton (Clark 1998).

Interseeding Cover Crops

Greater amounts of OM are produced if the cover crop is sown before the primary crop is harvested. Extra growth of the cover crop can be obtained by seeding it in the last cultivation or just before the beginning of leaf drop. Interseeding by air is a low-cost method of starting a number of cover crops and works well if there is sufficient moisture for germination and early seedling growth or if there are enough residues, either from the current crop or previous crops (Frye et al. 1988). The benefits are accentuated because of an early start at a time when there is little or no competition with the cash crop. Yet the period of growth is extended, allowing for the production of additional OM and increased fixation of N by legumes.

Reseeding Cover Crops

The cost of establishing a cover crop is usually low but can be reduced further by natural reseeding. Replanting the cover crop can be avoided by allowing viable seed to be produced on at least a portion of the cover crop. Waiting until the entire cover crop produces seed can shorten the season for the following crop and also increase costs of handling the cover crop. Mature nonlegume cover crops can limit yields of the succeeding crop due to wide C:N ratios, and both legume and nonlegume crops grown to maturity can deplete enough of the soil moisture to create some problems for the succeeding crop. Enough seed for reseeding the cover crop can be produced by leaving strips of the cover crop between rows to produce seed, and then killing the cover crop strips.

Kinds of Cover Crops

Eighteen crops are listed in the handbook *Managing Cover Crops Profitably* (Clark 1998) as suitable cover crops, rating which are most suitable for 14 different regions of the United States from the standpoint of N source, soil builder, erosion fighter, subsoil loosener, weed fighter, and pest fighter. The different cover crops are also rated as to amounts of dry matter produced and, for legumes, the amounts of N fixed, as well as their ability to scavenge residual N, their rapidity of growth, their forage or harvest value, their pH preferences, whether they can readily be interseeded in the cash crop and whether they reseed themselves, the lasting value of their residues, longevity, hardiness, and tolerances to heat, drought, shade, flood, and poor fertility. Additional valuable information is provided as to their type of growth habit, minimum germination temperature, seeding rates, depth of planting, cost of seed and planting, and type of inoculation needed for the legumes.

Potential advantages and disadvantages of each cover crop are also listed. The potential advantages covered are (1) impact of each crop on the soil from the standpoint of subsoiling, freeing of P and K, and loosening of topsoil; (2) soil ecology relating to the effectiveness of suppressing nematodes, diseases, or weeds and whether the crop has allelopathic properties; and (3) miscellaneous ratings as to the ability to attract beneficials, bear up under traffic, and fit in short windows between crops. The disadvantages include increased pest risks of weeds, insects, nematodes, and crop diseases, and whether the crop presents management challenges such as establishing the cover crop, hindering of the cash crop, difficulty of incorporating the mature cover crop, or difficulty of killing the cover crop by tillage or mowing.

Mixtures of Cover Crops

Often using mixtures of cover crops has advantages compared to growing any one cover crop. Mixtures of legumes with nonlegumes are rather common, as the benefits of N fixation combine with greater production of dry matter with some nonlegumes. Generally, because of their narrower C:N ratio, leguminous cover crops tend to break down and release N and other nutrients quicker than the nonleguminous grasses. Although this is usually a quick plus for the succeeding crop, it may have some disadvantages from the standpoint of weed control and building of SOM, which are usually poorer for the legume than the nonlegume. As was pointed out previously, combining legumes with grasses tends to give more favorable nutrient availability, biomass production, erosion control, and N scavenging.

Other advantages can be gained by using a mixture. Some of these advantages also apply to mixtures of legumes or nonlegumes. Common additional advantages sought are better nurse crops, breaking pest cycles, extending weed control, providing drought tolerance, reseeding, soil conditioning, and attracting and maintaining beneficial insects and mites.

Some of the characteristics of the more common cover crops and common mixtures of them are presented here. Much of this information derives from *Managing Cover Crops Profitably* (Clark 1998).

Crimson clover. Crimson clover, used as a winter annual in the Southern and Pacific coast states or as a summer annual in the northern states, is an efficient legume, fixing large quantities of N (75-150 lb N per acre) and is very useful for a wide range of farming operations. It has been used for strip tilling by killing 25-85 percent of the crimson clover with herbicides and planting corn. Sweet potatoes and peanuts have been successfully no-tilled into killed crimson clover. In northern climates, such as in Michigan, it is possible to successfully establish crimson clover after short-season crops, such as beans. In Maine, spring-seeded crimson clover is used for fall vegetables.

Crimson clover can be mixed with hairy vetch and nonleguminous cover crops, such as rye or barley. In mixtures, crimson clover is sowed at about two-thirds its normal rate, while the other crop is used at one-third or one-half its monoculture rate. Mixtures of crimson clover and hairy vetch attract beneficial insects and suppress weeds in Oklahoma pecan groves. In Ohio, a mixture of crimson clover, vetch, rye, and barley is being used as a mulch for no-till tomatoes. By combining several cover crops, a long-lasting residue is produced that permits reduction in herbicides, insecticides, and fungicides.

Hairy vetch. Hairy vetch, also a summer or winter annual legume, is outstanding as an N provider (90-200 lb/acre), weed suppressor, and topsoil conditioner. It can be interseeded in the primary crop in late summer before the primary crop is harvested, and it grows well with ridge tillage.

It can be grown in most of the United States, but in the deep South it is inferior to crimson clover. A form of hairy vetch that is almost smooth does better in the South as a winter cover crop, while the very pubescent type is more winter-hardy and does better in the North.

Its rapid spring growth keeps weeds suppressed, but its rapid decomposition limits weed suppression to only a few weeks. Much longer weed control (about five to six weeks) can be obtained by combining 30 lb hairy vetch with 10 lb crimson clover and 30 lb rye per acre. The mixture forms an excellent mulch for no-till vegetables, not only because of its ability to suppress weeds but also because of substantial N production and considerable enhancement of soil structure. Mixtures of hairy vetch with winter cereal grains of rye, wheat, or oats also fix substantial amounts of N and provide weed control. On the lighter soils of the coastal plain in New Jersey, Delaware, Maryland, and Virginia, mixtures of hairy vetch and small grains provide good wind erosion control as well as other cited benefits.

Rye. Rye is a cool-season annual that is suitable for most areas of the United States. It is probably the best cover crop for extreme winter temperatures and low fertility. Growth can be expected at temperatures not suited for legumes, and it will provide some cover where most crops will afford no protection. Rye, being a nonlegume, does not fix any N, but is a great scavenger of residual N and will provide longer-lasting OM.

Its ability to grow in cool weather makes it possible to plant rye later than most other cover crops and still gain some biomass, while obtaining considerable erosion and weed control. Rye reduces winter wind erosion, and by improving soil structure tends to reduce runoff, conserving water.

It has allelopathic properties against some weeds, making it desirable for no-till operations. These allelopathic properties are extended if the rye is left on the surface as a mulch, but this can be a problem for small-seeded crops, such as carrots or onions, planted in the mulch. The allelopathic effect is worse if the rye is killed when young. It is usually better to wait three to four weeks after terminating the rye before planting small-seeded crops. Larger-seeded crops and vegetable transplants are much less sensitive to the allelochemicals.

Combining rye with legumes, such as crimson clover or vetch, enhances the N available for the succeeding crop, especially if the rye is allowed to approach maturity before being terminated. The greater available N from the legume provides for early growth of the crop. In some areas of the country, rye and hairy vetch are interplanted, providing high N fixation with longer-life OM.

Rye and vetch mixtures have been used successfully for many years as cover crops before planting vegetables. Growing rye with the legume allows growing the legumes for a few weeks longer, thus increasing the yield of biomass and N. The mixture with vetch also provides better weed control. Vetch tends to climb on the rye, obtaining more light and fixing additional N.

When planted with a legume, rye needs to be planted at the lowest recommended rate for that area, but at low or medium rates if planted with grasses. Mixing 42 lb of rye with 19 lb of hairy vetch per acre was optimum for fall seeding before no-till corn in Maryland. A higher rate of 56 lb/acre was preferred if the rye was seeded with clovers. For transplanting tomatoes into hilly soil subject to erosion, a mixture of 30 lb rye, 25 lb hairy vetch, and 10 lb crimson clover has provided good results, evidently by providing excellent N, biomass, and improved soil conditions.

Ryegrass. Annual ryegrass, also known as Italian ryegrass, can be grown as a cool-season annual in most areas of the United States. It does well with good moisture and fair fertility but can be established quickly even in rocky or wet soil and will survive some flooding once established. It is a great erosion fighter and enhances soil structure, increasing water infiltration and water-holding capacity. It establishes itself very quickly, providing good weed control and scavenging for N.

Although considered as an annual, ryegrass may become a biennial in some regions, such as the mid-Atlantic states. Here, some plants can overwinter, growing quickly and producing seed in late spring, which can lead to a weed problem. In such cases, the crop needs to be terminated before seed forms.

Ryegrass mixes well with other grasses and legumes. It can be combined with red and white clover and overseeded in corn at last cultivation or in pepper, tomatoes, and eggplants when plants are in early or full bloom. Rates used are 15-30 lb of ryegrass and 5-10 lb of the clovers per acre. If aerially seeded, rates need to be increased by about 30 percent.

Several other methods of seeding mixtures are suggested. Mixtures of ryegrass and legumes or small grains can be planted in the fall or early spring, using 8-15 lb of ryegrass and two-thirds of the usual rate of legume. A mixture of equal parts ryegrass and crimson clover does well in certain areas of California. Frost seeding of red clover or other cool-season legumes with large seeds can also be used. (Frost seeding utilizes alternate freezing and thawing to move seed scattered on the surface deep enough to germinate in favorable weather without any tillage.)

Barley. Barley, a nonlegume, is adapted to most areas in the United States as a cool-season annual. It is very easy to grow but prefers dry, cool areas. It can be grown as a spring cover crop further north and produces more biomass than any other cereal. In warmer areas of the country, it can be grown as an overwintering cover that develops a deep root system and provides excellent erosion control. The rooting system as a spring crop is much shallower but still provides good erosion control. Its large production of biomass plus its strong root system improves soil structure and water infiltration. Because of its rapid start, quick shading of competing plants, and its content of allelochemicals, it is an excellent weed suppressor.

Barley does well when mixed with different grasses or legumes. A mixture of oat, barley, and a short-season field pea ('Trapper' variety) has been suggested for Vermont. Barley is added to a mix of brome, fescue, lana vetch, and crimson, red, and subterranean clovers at 10-20 percent of the mix, and the mix is seeded at a rate of 30-35 lb/acre for northern California vineyards. Barley can replace rye in some mixes but generally does not do as well as rye unless the weather is hot.

Wheat. Winter wheat can be grown as a winter annual in most areas of the United States. It has advantages of helping control weeds and resisting erosion. It improves soil by adding biomass, although amounts produced are usually smaller than those produced by barley or rye. It is easier to kill than barley or rye, making it less likely to become a weed. It also tolerates wetter soils than barley or rye but is easily killed by flooding. It is slower to mature than barley or rye, providing extra time for incorporation in the spring.

Several mixes of wheat and legumes are practical. Wheat is a good nurse crop for frost seeding red clover or sweet clover. The red clover is seeded at 8-12 lb and the sweet clover at 4-12 lb/acre. Wheat can be used in place of rye in legume mixes but tends to be less cold- and drought-tolerant than rye.

Field pea. Field pea is a legume that can be grown as a summer annual in the northeast, north central, and northern plains and as a winter annual in the southern states. There are two types of field peas. Austrian winter peas are sensitive to heat and humidity, failing to set seed at temperatures above 90°F and will do poorly if temperatures exceed 86°F. Where adapted, as in North Carolina, fall-seeded Austrian peas will outyield hairy vetch, common vetch, or crimson clover. Canadian field peas are related to the Austrian peas but have larger seeds, requiring higher seed rates. The Austrian field pea is most commonly used and is the pea referred to here.

Field peas, where adapted, are good producers of biomass and N, and are more efficient in using water than lentils or black medic. They break down rather quickly with quick release of N. The rapid breakdown makes them unsuitable for use as a mulch for no-tilled crops. They provide good disease suppression in dryland operations and can break cycles of some diseases, such as *Septoria* leaf spot, but are susceptible to ascochyta blight, fusarium, and *sclerotinia trifoliorum* eriks. Diseases can be minimized by rotating cover crops and avoiding successive crops of field peas.

Several mixes of peas with other cover crops are used. Field peas are mixed with small grains for dryland forage. The mix helps trap snow and provides early spring moisture. Mixed with rye, wheat, or spring oat, peas produced more biomass and N than mixtures of hairy vetch, common vetch, or crimson clover with these cereals when planted in coastal plain soils of North Carolina.

A mixture of field peas and rye is used as a mulch for growing tomatoes. In an Ohio study, evaluating three different ratios of rye: field pea seeds (3:1, 1:1, and 1:3) and three different seeding rates for each ratio, it was found that the highest rates of the 1:1 and 1:3 ratios gave the best yields of tomatoes. The cover crop was mowed 61 to 74 days after planting, and one-month old tomato transplants were put in 2 to 12 days later. The highest rate of the 1:1 ratio equals about 51 lb rye: 63 lb field peas per acre and that of the 1:3 ratio equals about 25 lb rye: 95 lb field peas per acre (Akemo et al. 2000).

Warm Weather Cover Crops

Several warm weather cover crops can be grown in the summer between primary crops. The leading candidates for these summer windows in temperate zones and at various times in the semitropical and tropical regions are cowpeas, velvet beans, buckwheat, and sorghum-sudangrass hybrids. Cowpeas and velvet beans are legumes; buckwheat and sudangrass-sorghum hybrids are nonlegumes.

Cowpea. Cowpea, also known as southern, blackeye, or crowder pea, is one of the most heat-tolerant legumes. It does very well in moist areas but tolerates drought and low fertility, while providing good weed and erosion control, and fixes substantial amounts of N (100-1,250 lb/acre). It attracts beneficial insects, such as lady beetles, honeybees, several types of wasps, and soft-winged flower beetles.

A rapid growing mix of 15 lb of cowpeas and 30 lb of buckwheat per acre provides a good cover crop in a period of about six weeks. Replacing about 10 percent of the cowpeas with sorgum-sudangrass hybrids increases biomass production, particularly during dry periods. The upright sorgum-sudangrass hybrid helps support the cowpeas during harvest, making them easier to mow. Cowpeas can also be mixed with pearl millet, which may do better on sandy soils.

Velvet bean. Velvet bean, also known as mucuna, is a legume that is little used in the United States, but has had some acceptance in Central America. Its tolerance to heat and moisture makes it a possible candidate for long open windows between crops during the summer in moist regions of the United States and many areas of the tropics. Its relatively low N production as compared with a sorghum-sudangrass hybrid (Table 6.1) in a short growing season reduces its potential use. Better N production can be expected if planted for longer periods.

Little has been done in evaluating mixes of velvet beans with other crops, but its performance under tropical, long-season, hot climates makes it an interesting candidate for mixed seedings in these areas, when combined with nonlegumes of about equal height.

Buckwheat. Buckwheat is a summer or cool-season annual that makes very rapid growth and thus is often used in windows between crops. Although grown in the summer, it does not do well under hot, dry conditions nor in compact dry soils or overly wet soils. Its tolerance of poor fertility and its rapid growth make it a good candidate for poor or weedy soils. It breaks down quickly, releasing nutrients for the succeeding crop. It harbors many beneficial insects, such as hover flies, wasps, minute prairie bugs, insidious flower bugs, tachnid flies, and lady beetles.

Its rapid growth limits its use as a nurse crop, but it can be combined with late-planted legumes because it will be killed by freezing temperatures, allowing the legume to thrive. Mixtures with sorghum-sudangrass hybrids are also useful.

Sorghum-sudangrass hybrids. Sorghum-sudangrass hybrids are summer annual grasses that thrive in warm temperatures of most areas of the United States. It is tall and fast growing, enabling it to quickly smother weeds. If mowed once, root systems tend to open compacted soils. It tolerates soils with low fertility and moderate acidity and alkalinity, but will do better in soils of good fertility and near-neutral pH.

Sorghum-sudangrass does well in mixes with other warm-season plants. Mixes with a nonlegume (buckwheat) or legumes (sesbania, sunn hemp, forage soybeans, or cowpeas) produce excellent cover crops. The buckwheat would have an advantage in weedy fields because its fast growth can be expected to suppress the weeds. When mixing with sunn hemp, also known as crotalaria, care should be taken to match the heights of the two plants. The sorghum-sudangrass supports the other legumes and there should not be a problem of the companion crop shading the sorghum-sudangrass.

Mixtures of sorghum-sudangrass in a 79:21 ratio with cowpeas, a 96:4 ratio with soybeans, or a 99:1 ratio with velvet beans provided large amounts of biomass with good weed control as summer cover crops in North Carolina. The sorghum-sudangrass/cowpea mixture gave the best weed control, but the sorghum-sudangrass/soybean mixture provided the most biomass (Creamer and Baldwin 2000).

Tropical kudzu. Tropical kudzu has been used very effectively in restoring fertility of worn-out tropical forest soils. After eight months of growth, tropical kudzu provided corn yields equal to that obtained with ten years of traditional fallow. Weeds, which normally overrun fallow soils, were suppressed by the tropical kudzu. Tropical kudzu, because it does not have the storage roots of the temperate kudzu, is easy to eradicate by burning or slashing before incorporating in the soil (Tropical kudzu 1988).

PLANT RESIDUES

Crop residues are the major sources of OM in many areas. Roots, stover, leaves, stems, orchard trimmings, and other plant parts contribute to the total. For many crops, the roots are the major part of the residue, but they and other parts are increased as inputs of lime, fertilizers, etc. are maximized for the crop. As has been repeatedly pointed out, burning residues to ease preparation of the next crop or for insect and disease control needs to be discouraged. Burning destroys a major part of OM with a simultaneous loss of the volatile elements N and S, and should be avoided unless absolutely necessary for pest control. Often, shredding the plant material before incorporation, adding N to speed its decomposition, and allowing enough time between incorporation and planting will accomplish the same purpose as burning without the attending losses.

Removing straw residues from grain fields also needs to be discouraged. Such products have some value as animal feed, but are used primarily for bedding or diverse industrial uses. But their values often are so low that it would be more profitable in the long run to allow them to improve SOM.

The amounts of crop residues produced can be large. Amounts of residues have increased with better fertilization practices and the introduction of newer high-yielding varieties. Some high-yielding corn crops have produced as much as 3.5 tons of stover and roots per acre, and some double cropping systems have yielded almost 6 tons of residues per acre annually. The average amounts of residue produced by some high-yielding crops are given in Table 6.3.

Adding the residues left over in the form of roots or rhizomes would substantially increase the amounts of dry matter returned as residues. Unfortunately, reliable figures about the weight of below-ground residues are not readily available. Part of the problem is the difficulty of getting data about roots, especially for the entire profile, which may extend to a depth of several feet. Also, amounts vary tremendously depending upon the type of soil, amounts of nutrients, rainfall, and its distribution as well as the kind of crop. Several studies indicate large amounts are produced by perennial grasses—2,000 lb or more per acre in the upper foot of soil, while alfalfa produces about 6,000 lb/acre. Red clover produces more than white clover—about 2,500 versus 1,500 lb/acre. Winter cereals produce about 2,500 lb/acre and spring cereals about 1,500 lb/acre. All of these produce much more than a crop of potatoes, which usually creates less than 500 lb/acre.

TABLE 6.3. Amounts of residues produced by several crops

Crop	Residue	Amount (lb/acre)
Barley	straw	3,840
Corn	stover	9,000
Cotton	burs, leaves, stalks	6,000
Oats	straw	5,000
Peanuts	vines	5,000
Rice	straw	6,000
Sorghum	straw	7,500
Tobacco	stalks	3,600
Tomatoes	vines	4,500
Sugar beets	tops	32,000
Wheat	straw	4,500

Source: Adapted from Wolf, B., J. Fleming, and J. Batchelor. 1980. *Liquid Fertilizer Manual.* Peoria, IL: National Fertilizer Solutions Association.

Residues Produced with Reduced Tillage

Amounts of residue increase substantially as yields of the primary crop increase and if the crop is grown in continuous no-till as compared to conventional till. In Illinois, the residue from prior crops tended to build up under continuous no-till systems, adding about 7 percent surface cover for nonfragile crops such as corn and about 3 percent for the fragile soybean crop (Table 6.4). For a list of common fragile and nonfragile plant residues, see Table 2.1.

Saving the residues aids SOM maintenance but in many cases fails to add enough OM to increase SOM. In most cases, additional residues can be helpful. Generally, the amounts of residues can be increased by good fertility practices, especially as enough N is introduced in the system.

Variable Effects of Different Residues

There often are marked differences in the benefits of different residues. For example, wheat straw is better in preserving infiltration than German millet, which in turn is better than sorghum residues. Some of the benefits may relate to differences in weight produced by various crops but often relate to other factors, such as different densi-

TABLE 6.4. The influence of corn and soybean yields upon percent surface covered by residue after harvest

Crop	Yield (bu/acre)	Surface cover	
		Conventional till (%)	Continuous no-till (%)
Corn	<100	80	87
	100-150	90	97
	>150	95	100
Soybeans	<30	65	68
	30-50	75	78
	>50	85	88

Source: Farnsworth, R. L., E. Giles, R. W. Frazes, and D. Peterson. 1993. *The Residue Dimension.* Land and Water # 9. Cooperative Extension Service, University of Illinois at Urbana-Champaign.

ties and diameters of the residue, how they are handled, their ability to last over longer periods, and allelochemical effects.

The extent of surface coverage after harvest is an important factor in determining the effectiveness of the residue, and it is affected by the type of crop, its yield, how the crop is harvested, and subsequent cultivation. The importance of using residues as mulch and various factors affecting the efficiency of mulches are covered in Chapter 8.

Handling Crop Residues for Pest Control

Crop residues, especially roots and stems, can harbor disease organisms, damaging insects or parasitic nematodes for some time, and often are the source of infection for the next crop. The manner in which these crop residues are handled often determines the extent and severity of the infection for the next crop.

Burning and deep plowing residues have helped reduce the amount of infection and have been recommended as a means of control. Such recommendations have been followed to control such diverse diseases as diplodia disease of corn, caused by *Diplodia zeae* (Schw.) Lev, and Texas root rot of cotton caused by *Phymatotrichum omnivorum* (Shear) Duggar. Burning of grass seed fields in Oregon has been used for many years to control several fungal diseases as well as killing weed seeds and destroying mites and insects. By deep plowing roots out of the soil and disking as soon as possible in order to promote decay of residues, the North Carolina R-6-P program reduces damage from brown spot and mosaic in tobacco as well as controlling tobacco hornworms (*Manduca sexta* L.), and tobacco budworms (*Heliothis virescens* Fabricus).

Burning residues, however, leads to losses of N and S and most of the other benefits that accrue from OM. Almost all of the benefits usually gained from OM, such as improved physical effects and the stimulation of biological processes, are eliminated. Burning of residues results in the loss of SOM as well as OM. Some loss of SOM near the surface occurs during the combustion process, and some of it may occur later because of the accelerated erosion that often follows burning.

Accelerated erosion is quite common if rains fall soon after burning, as burning often reduces infiltration. The reduced infiltration appears to be the result of lower SOM and of certain chemicals produced in the combustion process that are deposited in soil pores.

Some OM benefits are also lost by the deep plowing of residues, although it too has been quite effective in controlling several plant diseases and insect pests. Deep plowing eliminates any benefits of keeping plant residues at the surface, including lower soil temperatures during heat of the day, better water infiltration, lessened soil erosion, and lessened soil compaction from rains and irrigation.

Avoiding burning or deep plowing of residues. Serious attention is being given to find alternative methods to burning and deep plowing. It is now generally recognized that proper use of residues for most farming operations must be encouraged if we are to maintain a sustainable agriculture.

At times, it may be possible to reduce disease or insect pressures by shredding the harvested crop and incorporating it shallowly into the soil a few weeks before planting the next crop. This hastens decomposition of the residues and reduces the chances of the disease and some insect organisms affecting the next crop. Although this eliminates some of the benefits of keeping the residues at the surface and may expose the soil to erosion, it does retain some benefits, such as improved porosity and infiltration.

It also may be possible to do away with the practices of burning or deep plowing for pest control by using one or more of the following:

1. Resistant cultivars
2. Better sanitation
3. Obtaining better knowledge of the life cycles of the disease organisms
4. Using rotations
5. Adding organic matter
6. Reducing tillage

Resistant cultivars have already been developed that make it possible to produce good yields of many different crops despite the presence of large numbers of disease-causing organisms.

Better sanitation may consist of employing such practices as use of disease-free seed, disinfecting contaminated seed, seed treatments, fungicidal drenches applied while planting and/or during the life of the crop, cleaning debris from harvesting equipment, or sanitizing harvesting and other equipment prior to moving into new fields.

Better knowledge of diseases and insect pests has already led to some changes that have produced improved control and promises to further reduce problems in the future. Changes in planting dates and cultivation practices that favor the crop or put the disease organisms or insects at a disadvantage have already been implemented. In some cases, they have also led to the development of rotations that allow the crop to be grown with minimal treatments because the rotation without susceptible crops greatly reduces pest pressure. More about crop rotations is presented in Chapters 2, 9, 10, and 11.

Reduced tillage appears to give better control of certain pests for several reasons. Probably the most important is that it promotes increases in SOM while utilizing the OM to better advantage as it is kept as mulch. More about reduced tillage is presented in Chapter 9.

Often a combination of practices provides the best pest control. For example, a marked reduction in fungicide use for several vegetables (cabbage, cantaloupes, cucumbers, eggplants, lima beans, snap beans, southernpeas) and peanuts grown in Georgia has been made possible by increasing organic matter input and reducing tillage, which maintains a very effective mulch. In addition to good sanitary practices and rotation of crops, the system adds more organic matter with the use of crimson clover and winter rye. Crops are planted in the killed cover crop left on the surface as mulch. Strip tilling of the rye has also been used to add organic matter and keep tillage at a minimum. A good part of the control reported for Georgia appears to be derived from the additional OM introduced with greater use of cover crops and reduced tillage (Phatak 1998).

Chapter 7

Adding Organic Matter Not Grown in Place

Although it is usually cheaper to provide OM by growing it in place, some other OM can be obtained at reasonable costs. What is reasonable is influenced greatly by the sale price of the crop produced and also by the nature of the OM and its cost, including transportation and handling.

Producers of high-priced agricultural products, such as nursery items, herbs, cut flowers, fruits, or vegetables, usually can afford to pay more for OM inputs than producers of milk, beef, sugar, and small grains. The producer of specialty items and those who have special markets or are covered by certain governmental programs may find the cost of certain forms of OM well within acceptable ranges, whereas these items may be prohibitive for general producers.

Transportation and handling costs make a number of OM materials cost prohibitive for many farmers. Items such as manure or composts produced on the farm, even though they are not produced in place, may have an advantage because of low transportation costs. These items become even more attractive economically if their handling can be mechanized.

Organic matter not produced on the farm generally has been too costly for most growers, but this may be changing. The recent trend of free OM in the form of composts, manure, and sewage sludge (biosolids) resulting from pollution control measures has changed the economics of a number of organic materials. The lack of sufficient landfill sites has prompted a number of cities to compost certain organic wastes. Sometimes produced in amounts greater than can be used on city parks and recreational areas, it often is available to the public at no cost at the site. A somewhat similar situation may be developing with biosolids as disposal of human wastes into bodies of water comes under more stringent regulation. The regulation by states and the federal government regarding disposal of animal manure also is making more of this item available at no or low cost.

MANURE

Animal manures have been used for centuries as a basic source of plant nutrients. Besides supplying nutrients, manure adds considerable OM, improves soil structure, increases moisture holding capacity, and provides for a high degree of biological diversity in the soil. Up until about 75 years ago, it was an integral component of farming, allowing an orderly disposal of a waste product from animal-dependent agriculture that encouraged satisfactory crop yields with a minimum of costly inputs.

Routine manure use began to change after World War I as horses gave way to tractors, automobiles, and trucks. A combination of using smaller amounts of manure or none at all, elimination of large acreage devoted to hay and pasture, and the purification of fertilizers, which decreased micronutrient content, led to serious micronutrient shortages beginning in the 1930s and lasting until about the 1960s. The problem of micronutrient shortages has been largely overcome by their general addition to fertilizers, but their cost can largely be eliminated by the use of manure.

The use of manure today usually does not make the economic sense of former years. Manure still can be an important source of nutrients in regions close to its source, but relatively low analyses make it expensive if it has to be hauled appreciable distances. Instead of being produced on farms with sufficient crop acres to use the manure profitably, much of it is now produced in concentrated areas, such as cattle feedlots and pork, broiler, or egg factories (see Table 7.1). Many of these intensified production units are far from cropland that can use the manure efficiently. Even if cropland is available, such large amounts of manure are produced that if all of it was economically disposed of on nearby cropland, it would exceed tolerable application levels, probably causing poor yields due to excessive salts or nutrients, with a potential for groundwater pollution.

As will be seen, high additions of water reduce nutrient concentrations in manure and make the product less attractive from a nutrient standpoint, yet moving manure by water may make handling large

TABLE 7.1. Annual production of manure in the United States by confined animals and their contents of N, P_2O_5, and K_2O (million tons per year)

Animal manure	Total production	N	P_2O_5*	K_2O**
Poultry	54.0	0.7	0.45	0.30
Swine	15.5	0.73	0.42	0.73
Beef	24	0.51	0.41	0.63
Dairy	24	0.78	0.30	0.63

Source: Based on data presented by Edwards, J. H. and A. V. Someshwar. 2000. Chemical, physical and biological characteristics of agricultural and forest by-products for land application. In *Land Application of Agricultural, Industrial and Municipal By-Products*. Madison, WI: Soil Science Society of America, Inc. Reproduced by permission of the Soil Science Society of America, Madison, WI.
*To obtain amounts of P, divide by 2.2914.
**To obtain amounts of K, divide by 1.2046.

amounts of it economically feasible. Moving manure by water greatly reduces the cost of transportation and application, making the necessary disposition of farm manure less costly and somewhat more competitive with fertilizer.

Amounts

In 1981, it was estimated that the total manure produced annually by confined domestic livestock and poultry in the United States was about 990 million tons, but only about 10 percent was spread on cropland (Follett et al. 1981). A later estimate placed the total amount of manure by-products for confined animals at 159 million tons, of which about 61 million tons are collected annually (Edwards and Someshwar 2000).

Nutrient Value

Although the long-term effects of manure on SOM contributed to much of its value, it was largely the nutrient content that was prized by growers. Used at rates varying from about 10 to 20 tons fresh weight per acre, the application often supplied all the nutrients needed for the succeeding crop. As it became cheaper (in the short run) to buy nutrients in fertilizer than to obtain it from manure, means were sought for its disposal, and it often was dumped in rivers or streams. Government regulations prohibiting such dumping, which have stopped this procedure, probably helped increase manure's value, but it has largely been our recognition of the need for OM that has prompted us to take a new look at manure and other organic materials.

Manure comes off very well in this scrutiny because of its potential nutrient content and contribution to SOM. The nutrients potentially present can partially or totally substitute for fertilizers, depending upon the quantity and quality of the manure. The total nutrients available from different types of manure are tremendous (Table 7.1).

Variation in Nutrient Content

Unfortunately, the nutrients actually supplied by manure often do not meet their potential. The nutrient content can vary considerably from average values given in Table 3.2. Some of the variation is illustrated by an analysis of dairy and poultry manure from two states (Table 7.2).

Some of the variation in nutrient content is due to differences in content of feces, urine, and bedding, as these items vary considerably in their composition, but some of it is due to differences in rations or the amount of bedding and/or water added. Unfortunately, much of the low nutrient values are due to poor storage and handling of the manure (Table 7.3). The losses of N are most serious, but some P and K along with other elements are also lost—primarily as runoff or by leaching from open storage pits. Some settling out of nutrients in open lagoons can also contribute to losses. Application methods also affect N loss (Table 7.4). Soil incorporation reduces these losses.

TABLE 7.2. Variation in N, P_2O_5, and K_2O analyses of dairy and poultry manure collected from two states (lb/ton)

State	Dairy	Poultry	Dairy	Poultry	Dairy	Poultry
	N		P_2O_5		K_2O	
N. Carolina	4-12	15-48	2-9	14-30	2-13	7-16
Maryland	4-14	4-136	2-9	10-111	1-15	1-79

Source: Livestock Manure Folder. Potash and Phosphate Institute and Foundation for Agronomic Research, Norcross, GA. Reproduced by permission of the Potash and Phosphate Institute, Atlanta, GA.

TABLE 7.3. Losses of N, P, and K from manure as affected by handling and storage*

Handling and storage	Losses of nutrients (%)		
	N	P	K
Solid systems			
Daily scrape and haul	15-35	10-20	20-30
Manure pack	20-40	5-10	5-10
Paved lot	40-60	20-40	20-30
Deep pit (poultry)	25-50	5-15	5-15
Litter (poultry)	25-50	5-15	5-15
Liquid systems			
Anaerobic deep pit	15-30	5-15	5-15
Aboveground storage	5-25	5-15	5-15
Earthen storage pit	20-40	10-20	10-20
Anaerobic lagoon	70-85	50-85	50-75

Source: Sutton, A. L., D. W. Nelson, D. D. Jones, B. C. Jones, and D. M. Huber. 1994. *Animal Manure As a Plant Nutrient Source.* Purdue University Cooperative Extension Service ID-101, Purdue University, West Lafayette, IN.
*Based on composition of land-applied versus that of freshly excreted manure and adjusted for dilution effects of different systems.

TABLE 7.4. Loss of nitrogen from manure as affected by application method

Application method	Type of manure	N loss (%)
Broadcast, not incorporated	solid	15-30
	liquid	10-25
Broadcast, incorporated	solid	1-5
	liquid	1-5
Injected	liquid	0-3
Irrigation	liquid	30-40

Source: Sutton, A. L., D. W. Nelson, D. D. Jones, B. C. Jones, and D. M. Huber. 1994. *Animal Manure As a Plant Nutrient Source.* Purdue University Cooperative Extension Service ID-101, Purdue University, West Lafayette, IN.

Some of the N loss can be reduced by the addition of superphosphate (0-20-0) to the manure. Superphosphate, mixed with the manure at the rate of 25-30 lb/ton, helps fix N that may be lost as ammonia (NH_3) in a nonvolatile form, probably as ammoniated superphosphate. It also has the added advantage of supplying P, which is often lacking in some manure. Additional N and much of the K_2O can be saved by (1) using bedding to reduce loss of liquid excrement, (2) packing the manure tightly in lined pits with no drainage, (3) storing it under cover, or (4) applying it to the field and working it into the soil as soon as possible.

It is evident that considerably more N and other elements would be available for crop production, and there would be less pollution of water sources, if we did a better job of handling, storing, and applying manure. The problem of nutrient loss is not as serious on farms where animals and crops are combined into functioning units. On such farms, much of the manure is deposited in pastures, and the remainder from confined animals can be readily moved almost on a daily basis to crop lands or stored for short periods under cover. More serious problems exist in feeding lots and animal "factories," where large amounts of manure cannot readily be moved to crop fields. As a result, the manure is stored in packed piles, aerated lagoons, or anaerobic lagoons. The loss of nutrients from stored packs may not be much different than from daily scrape. Aerated lagoons (either shallow enough to allow ready penetration of air or with air introduced) have less loss of nutrients than the daily scrape and haul method. The greatest loss (about 75 percent of the nutrients) occurs with the anaerobic lagoon.

Considerable variation in nutrient content is also due to variable moisture contents. Unless it has been dried or composted, manure tends to have considerable moisture, varying from about 30 to 80 percent. Most manures tend to be relatively low-analysis materials. Even on a dry weight basis, they seldom analyze more than 5 percent N, 2 percent P_2O_5, and 3 percent K_2O. The high levels of moisture reduce the nutrient content to a point that there is little incentive to use them, considering that 100 lb of a dry fertilizer analyzing about 15-15-15 can supply 5 to 15 times the nutrient content of an equal weight of manure. The problem is compounded as manure is stored in lagoons or pits, but these can readily be moved mechanically by irrigation systems, making their cost of handling negligible.

The value of manure is due to much more than its nutrient content. Its effect on soil structure, biological diversity in the soil, and as a source of SOM may outweigh its value as a nutrient source, but it is important to retain as much nutrient content as possible. The necessary disposal of manure makes it desirable to avoid nutrient waste, making disposal valuable. For manure produced some distance from cropland suitable for its disposal, even the best methods of handling and storage may not make the product competitive for use on cropland. Making changes in the manure, such as drying and forming it into pellets or composting, could make it suitable for home garden use, ornamental crop production, or landscape purposes.

Source of Micronutrients

In addition to N, P_2O_5, and K_2O, animal manure can be a good source of other macronutrients (Ca, Mg, S) and most essential micronutrients. Micronutrients present in manure and the chelation of these elements by various manure fractions help to prevent micronutrient deficiencies, and probably explain why there are few known cases of micronutrient deficiencies where large amounts of manure are routinely used for crops on acid soils. Unfortunately, the micronutrients supplied by manure are of less value on high-pH soils. The large quantities of CO_2 produced by decomposing manure in combination with bicarbonates usually present in these soils tend to maintain or increase pH, reducing micronutrient solubility. The reactions taking place can be illustrated as follows:

- Carbon dioxide (CO2) formed from microbial and root respiration reacts with water to form carbonic acid.

$$CO_2 + H_2O \leftrightarrow H_2CO_3$$

- In high-pH soils, the abundant OH^- ions tend to form bicarbonate ions (HCO_3^-) but this is followed by the formation of carbonate ions (CO_3^{2-}). An increase in CO_2 shifts the reaction to the left, which increases the amount of OH^- ions and increases the pH.

$$H_2CO_3 + OH^- \leftrightarrow HCO_3^- + H_2O$$
$$HCO_3^- + OH^- \leftrightarrow CO_2^{2-} + H_2O$$

- As illustrated in the final reaction for the micronutrient Fe, abundant OH^- ions at the high pH alters the simple soluble Fe^{3+} ion to soluble ferrous hydroxide and finally to the insoluble ferric hydroxide.

$$Fe^{3+} \xrightarrow{OH^-} Fe(OH)^{2+} \xrightarrow{OH^-} Fe(OH)_2 \xrightarrow{OH^-} Fe(OH)_3$$

Nutrient Release

Manure can supply appreciable amounts of macronutrients and micronutrients, but the release of the nutrients depends on the type of manure, how it is handled, and also on soil factors as cited for release of nutrients from organic matter.

Because all nutrients in manure are not immediately available to the crop or even available in the first year, it is difficult to match nutrients supplied by manure to those needed by the primary crop. The release of P and K is relatively rapid, but since N may be immobilized by microorganisms, a good part of it may not be available in the first year, and all of it may not become available for several years. The slow release of N, while making it difficult to calculate N needs, is an asset since it provides for less N leaching and a more uniform availability than N from most inorganic sources.

Various methods have been used to calculate the amount of N that will be available. Although there is considerable variation in N release due to differences in manure and climatic conditions, a general estimate has evolved that only about one-third to one-half of the N can be expected to be released for the first crop year after application. Scientists at the University of Illinois suggest that 40 percent of the N is available the first year, 30 percent the second year, 20 percent the third year, and 10 percent the fourth year. The University of Maryland assumes only about a 5 lb N release per ton of manure, as it suggests a 50 lb reduction in N per acre if 10 tons of manure are applied for corn. Somewhat similar estimates have been made by Michigan State University in its recommendation for N use with and without manure applications.

Nitrogen release in the first year after application also can be estimated by calculating amounts of N mineralized in a year using a factor for different sources of manure and how they are handled. The amount of N available in the first year after application = (total N − NH_4-N) × mineralization factor + NH_4-N). The mineralization factor is 0.5 or higher (50-60 percent of the N is made available in one year) for fresh swine manure (0.50), deep pit poultry manure (0.60), and solid poultry manure with (0.50) or without litter (0.55). Most of the other manures have factors of 0.25-0.35, and horse manure with bedding has a factor of only 0.20.

The amounts released in the second, third, and fourth years would be approximately 50, 25, and 12.5 percent of the amount released the first year (Sutton et al. 1994).

It must be remembered that estimates for N release from manure and other organic sources are only that and, although helpful in calculating N needs of the crop, they often fail to give sufficient information to produce maximum economical yields. Since estimated N from OM may not be available at certain times (wide C:N ratios, inadequate temperature or moisture, rapid acceleration of microbial activity), it is necessary to add quick-acting N to make up the difference. Failure to do so, especially as the plant is getting established, can seriously reduce yields and/or quality. The need for additional N can easily be determined by soil tests for nitrate plus ammoniacal N, and sufficient quick-acting N can be added to provide for rapid growth.

Manure for Increasing SOM

Long-term application of manure can increase SOM. Soils at Rothamsted, England, receiving 14 tons/acre of farmyard manure annually for long periods had 5 percent SOM, whereas plots receiving only NPK fertilizers and cereal residues had only about 2 percent SOM. Somewhat similar results were obtained from long-term experiments at Saxmundham. Soils receiving 6 tons/acre farmyard manure had 3.5 percent SOM, and the NPK fertilizer plots had only 2.5 percent SOM (Cooke 1967).

In relatively shorter experiments, adding farmyard manure to a soil containing only 0.085 percent N (about 1.5 percent SOM) for a period of nine years at an annual rate of 15 tons/acre of farmyard manure resulted in a soil with 0.12 percent N (about 2.4 percent SOM).

The soil analyzed 0.14 percent N (about 2.8 percent SOM) after adding 30 tons of manure per acre annually. About 50 percent of the N in the applied manure was converted to soil organic N in SOM (Russell 1961).

COMPOST

Composting, or the formation of humuslike material outside the soil, also has been used effectively by growers for centuries. Composts have provided for good crop yields but have no advantage in crop production over the same materials added directly to the soil. In fact, adding the materials directly to the soil tends to provide more SOM and nutrients than if the materials are composted and then added to the soil. The reasons for using compost rather than applying the OM materials directly are a combination of convenience and the production of a superior product that is easier to handle. It often is not practical to add the materials directly because of timing, amounts of material to be disposed of, or the physical nature or location of the materials. By processing these materials as composts, they are preserved in a form that can easily be handled and stored, disposing of them at a time that is convenient for the grower, reducing the volume of materials almost by half, while providing a mixture that is fairly uniform and usually easier to handle than the individual materials. At the same time, the composting process, if carried out properly, will often reduce odor problems, nutrient loss, and noxious weeds; provide a better C:N ratio than the raw materials; render harmless many toxic substances associated with the materials, such as pesticides and allelochemicals; partially sterilize the materials; and provide a material that tends to suppress soilborne diseases.

Materials for Composts

Composts have been made from a wide variety of organic materials such as manure, leaves, grass clippings, bark, wood scraps, kitchen scraps, pine needles, straw, meat and fish scraps, and spoiled hay. In recent years, as municipalities have made composts to reduce the need for landfills, the materials used generally have been garbage, yard and garden clippings, biosolids, and wastes from produce or processing plants.

The wide range of materials used contributes to variability in the bulk density, particle size, water-holding capacity, and chemical composition of the finished compost. A wide variability of several important compost characteristics can be expected (Table 7.5), although not all variability is due to differences in original materials. Some may be the result of different methodologies for producing composts.

Some care needs to be taken in selecting materials in order to have a desirable product. A balance of wide C:N ratio materials, such as straw, wood chips, paper, and brown leaves, needs to be mixed with relatively narrow C:N ratio materials, such as manure, animal residues, grass clippings, green leaves, legume hay, or biosolids.

Adding excess quantities of wide C:N ratio materials will require frequent mixing and adjustment of moisture levels to provide a suitable mix. Excess use of narrow C:N materials makes it a bit more difficult to control odors and may lead to greater N loss. If enough narrow C:N ratio materials are not available, some fertilizer N can be added. A C:N ratio of about 30:1 or less needs to be obtained from the mixing of different materials and fertilizer N (if needed).

TABLE 7.5. Variability of several important composition criteria of compost

Test	Range of analyses
C:N	6:1-20:1
pH	5-8
Conductivity	0.2-2 S/m
Total N	0.5-3%
P	0.1-2.0%
K	0.2-1.0%
Ca	0.8-3.5%
Mg	0.3-0.6%
S	0.1-2.0%

Source: Based on data presented by Cooperband, L.R. 2000. Sustainable use of by-products in land management; and A.V. Barker, M. L. Stratton, and J.E. Rechcigl. 2000. Soil and by-product characteristics that impact the beneficial use of by-products. Both references in J.M. Bartels and W.A. Dick (Eds.), *Land Application of Agricultural, Industrial, and Municipal By-Products.* Soil Science Society of America Book Series #6. Madison, WI: Soil Science Society of America. Reproduced by permission of the Soil Science Society of America, Madison, WI.

Moisture and Oxygen Control

It is also necessary to control moisture and O_2 during composting. Ideally, the mix should have enough moisture to allow good microbial growth, feeling damp but not overly wet when squeezed. About 50-70 percent moisture is satisfactory. Aerating the mix by frequent turning or by introduction of air or exhausting it through pipes in the mix helps to dry out excessively moist mixes. Wood chips or other coarse materials in the mix aid air movement. Turning also helps provide a more uniform mix, moistens excessively dry areas in the pile, and prevents temperatures from rising too rapidly in early stages of decomposition (Figure 7.1).

Temperature Changes

If moisture and O_2 are maintained properly, microbial activity will increase, with a concomitant increase in temperature. In the first two

FIGURE 7.1. Aeromaster Compost Turner. A mechanical device for quickly turning compost materials piled in windrows. The frequent turning aids in providing uniform mixes with good oxygen supply and desirable moisture levels, controlling temperature, and providing rapid formation of high-quality compost. (Photo courtesy of Midwest Bio-Systems, Tampico, IL.)

to three days, when sugars and other readily biodegradable compounds are digested, the temperature will rise to the range of 104-125°F (40-51°C). Temperatures will continue to rise to about 158°F (70°C) but stay in a range of 125-150°F (51-66°C) for a few weeks as some of the more resistant celluloses are destroyed and some lignins are attacked. In a third or final stage also lasting several weeks, the compost cures as readily decomposable materials are exhausted. In this stage, temperatures drop to about 55°F (13°C), the compost is recolonized by mesophilic organisms from the outside of the pile, and humic substances accumulate. The final product will have a dark color and consist mostly of humic substances, lignins, and microbial biomass (Hoitink et al. 1996).

During the composting process, temperature is affected by ingredients, their particle size, moisture content, size of the compost pile, and ambient temperature values. High temperatures can be controlled by turning, which usually needs to be frequent at first but much less frequent as the compost ages.

Pile Size and Regulation of Oxygen, Moisture, and Temperature

Proper moisture and O_2 are aided by making compost piles the proper size. For small operations, pile size is usually limited to about 3-5 ft^2 and about 3-5 ft high. Such sizes can be turned and will retain fairly good internal moisture. The pile can be left unturned if the top is shaped to readily take in moisture, but such piles almost never provide a uniform product. Also, if left unturned, much of the exterior will not be heated enough to provide the disease suppression obtained when materials go through the full thermophilic changes.

Larger Operations

Larger farm or small municipality operations usually use windrows to produce compost, depending upon mechanical equipment to mix and move the materials. A typical compost operation, suitable for a small city or large farm operation, is conducted by the city of Douglas, Georgia, population about 15,000, producing about 8,000 tons of compost annually. It is making an excellent compost from two parts yard wastes (tree trimmings, grass clippings, dead plants) with one part biosolids (17-18 percent dry matter) recovered from treated sewage. Alternate layers of shredded yard wastes and biosolids are laid down in a windrow, which is repeatedly mixed by automatic equipment (Rick Reed, private communication, 2001).

Commercial or large-scale production of compost may be done in open systems using static piles or windrows, but in-vessel systems are becoming more popular, primarily to conserve space and time, but also to control objectionable odors. The static system of piles or windrows requires the least management time and is least expensive from an equipment standpoint. Static systems of aerated piles or extended aerated piles (windrows overlap, sharing aeration systems) can be easily established using commonly available equipment. Minimum needs are a well-drained surface, front-end loaders, and a power supply to move air, either by blowing or drawing it through perforated pipes located in the piles. Unfortunately, the static or open systems require a great deal of time (one to two years) and considerably more space. The in-vessel systems process a great deal of material in a short time span (7 to 14 days) but the quality of the compost is compromised. The lack of sufficient time at high temperatures produces a less stable material that lacks disease suppression, requiring additional curing to provide a high-quality product (Goutin 1998).

Need for Complete Process

The value of the compost is greatly increased if it goes through the entire process. The C:N ratio will drop from about 30:1 or more to a ratio closer to 10:1. The finished ratio is closer to that of microorganisms in the soil, eliminating any tie-up of N to meet the needs of the microorganisms. The finished product will offer some suppression of soilborne diseases and be relatively free of weeds and potentially toxic allelochemicals and pesticides. It is interesting to note that mature compost colonized by biological control agents (BCA) suppresses soilborne disease while the immature compost is of less value.

The complete composting process kills most beneficial and harmful organisms. The suppression of disease varies with different composts due to random recolonization as the compost is added to substrates. Suppression of *Rhizoctonia* spp. is more consistent if the compost is prepared in an environment high in many different types of microflora than if the compost is produced in an enclosed facility with limited microbial species. Compost prepared from municipal wastes that have been treated to kill pathogenic organisms will have less suppressive value than composts made from materials subject to less stringent treatments. The variability in suppression of *Rhizoctonia* spp. can be largely overcome by inoculating the nearly completed compost with specific microorganisms, such as the fungus *Trichoderma hamatum* and the bacterium *Flavobacterium balustinum 299T*, a process covered by a patent held by Ohio State University (Hoitink et al. 1996).

In addition to providing sufficient time to complete the compost processing, considerable attention must be paid to the C:N ratio of the materials, and temperature and moisture conditions prevailing at various stages of decomposition in order to produce a quality product. Douglas, Georgia, starts out with materials of satisfactory C:N ratio, and then pays a great deal of attention to temperature and *Escherichia coli* levels to produce a quality product. The *E. coli* levels are kept below 1,000 per gram by carefully monitoring temperature and turning the pile as per the following regime:

> First 10 days: Maintain at 100-145°F. Turn at least once daily.
> Days 10-35: Maintain at 145-165°F. Turn daily.
> Days 35-45: Maintain below 150°F. Turn as needed to keep temperature below 150°F.
> Days 45-60: Maintain below 140°F. Turn as needed to keep temperature below 140°F.
> Days 60+: Turn every two to three days until released from site (Rick Reed, private communication, 2001).

LOAMLESS COMPOSTS

Considerable compost is being used as a potting medium for vegetable and ornamental transplants and for growing ornamental plants. Composts for these purposes may be made with or without soil (loamless). The trend has been to prepare composts without soil because of the difficulty in obtaining soil free of herbicides or other pesticides. Although some pesticides can be broken down by a complete composting process, it is often safer to avoid soil that may contain pesticides. The preparation of loamless composts with ingredients low in

N yields a product that tends to be low in nutrients. Adding fertilizers during the decomposition process will largely overcome the deficiency. Inorganic fertilizers are satisfactory, but attention to conductivity must be paid to avoid excess salts. Use of relatively high-analysis slow-release organic materials make control easier, but coated fertilizers, such as Osmocote, can complicate the process as the coating can break down, releasing excess nutrients.

Storage of Loamless Composts

Because high-analysis organic materials or coated fertilizers can release considerable ammonia, their use requires attention to length of storage, particularly if the compost has not been allowed to go through its entire composting period. If used within the first week after preparation, loamless composts prepared with organic materials tend to produce about the same amount of growth as composts prepared with inorganic fertilizer. If compost prepared with organic fertilizer is stored for several weeks before using, it may produce very poor crops due to the ammonia produced as microorganisms attack the organic fertilizer. Ammonia is toxic in very small quantities, and the high pH that it produces tends to make several micronutrients unavailable. The ammonia will in time (six to eight weeks) be nitrified and the pH will drop, making satisfactory growth again possible. The exact time when this occurs varies with materials used and the moisture and temperature, but the process can be followed by measuring the pH at regular intervals. When the pH returns close to its starting point, most of the ammonia should have been nitrified.

Storing composts with slow-release inorganic materials can also be a problem, due to the salts, which can increase dramatically within a week or two. Materials such as Osmocote at elevated temperatures can release enough available nutrients in a relatively short period to make the composts unsatisfactory. Ideally, these composts should be used shortly after being prepared; if they must be stored, it is best to wait until the composting is complete, as indicated by a return to lower temperatures or a narrow C:N ratio. In many cases, it is more desirable to add the coated fertilizer after the compost is cured. A soil test taken after the compost is cured should be used as a guide. Add slow-release fertilizer only if salts are not already high, and use the mix shortly afterward.

Value for Crops

Numerous studies have evaluated composts for crop production. The readily available municipal composts have favored a great deal of experimentation. Most of these studies deal with ornamental, vegetable, and fruit crops, probably because the generally high value of these crops warrants considerable inputs. Results have been mixed.

It has been shown that composts made from municipal solid waste (MSW) can decrease soil bulk density and increase SOM, CEC, MHC, pH of acid soils, and both microbial and enzymatic action. As expected, vegetables and fruit trees have responded positively to compost applications, either as mulches or mixed with the soil. Plants were grown in pots, flats, or open fields. Soil and plant nutrient contents as well as yields have increased with additions of composts. In some cases, yield increases were maintained for more than one year. Compost mulch tends to reduce stresses caused by soil crusting and drought and has been useful for weed control. Reductions in blossom-end rot of peppers and losses to *Phytophthora capsici* have also been reported (Roe 1998). Studies indicate that standard compost made from wastewater sludge and yard trimmings is superior to those that may also contain mixed waste paper, refuse-derived fuel, or refuse-derived fuel residuals (Roe et al. 1997).

For the production of ornamental crops, compost has been successfully used as a potting medium, substituting for a mix consisting of six parts peat, four parts sawdust, and one part sand by volume. It has also done well for the production of ornamental trees when applied as a soil amendment in a tree nursery (Fitzpatrick et al. 1998).

Unfortunately, the positive effects of composts are not always present and, in some cases, there is an actual decrease in plant response. In many cases, early differences in germination or survival are not translated into yield increases. Comparisons with plastic mulches have indicated that compost mulches may not yield as well, at least some of the time. Comparisons of composts with fertilized plants indicate that plants receiving fertilizer may do as well. In some cases, best yields were obtained with both composts and fertilizer.

Negative effects usually appear to be related to the quality of the compost. Immaturity of the compost may be the major reason for poor plant response. The presence of allelochemicals and a wide C:N ratio in immature compost accounts for much of this poor response.

High pH values and/or excess salts in some composts have been a problem, especially for the production of ornamentals where the compost is used as the growing medium. Overly wet materials applied at high rates have contributed to some problems, probably by adversely affecting soil air. In some cases, high levels of heavy metals in the composts have affected plant growth adversely. Some composts contain inadequate N to support the fast growth of some crops, and composts made from sewage sludge or manure tend to cause deficiencies of Mn in some ornamental trees (Broschat n.d.).

It is obvious that special care must be exercised in selecting composts if favorable responses are to be obtained. Primarily, the compost needs to be aged properly to avoid problems of wide C:N ratios, allelochemicals, and possible toxic elements. A C:N ratio of about 15:1 or less probably would be a good indicator as to whether the materials have undergone the necessary reaction. Feel and odor can aid in selecting a compost that is uniform, not unduly wet, and with no unpleasant odor. Additional tests of pH, electrical conductivity, air space, water-holding capacity, and heavy metal content should help select composts that have greater chances for success. (Satisfactory values for these tests are 5.5-6.5 pH, 0.5-1.5 dSm^{-1} electrical conductivity, 5-30 percent air space, 20-60 percent MHC, As <75 ppm, Cd <85 ppm, Cr <3,000 ppm, Cu <4,000 ppm, Hg <840 ppm, Ni <75 ppm, Pb <420 ppm.)

Commercial Farm Use

Various advantages have been cited for compost use, such as improvement of CEC, pH, water retention, soil structure, SOM, and disease suppression with a decrease in fertilizer need and damage from soil contaminants. But because of its relatively low analysis (Table 7.5), it would appear that composts would be limited to restricted areas and uses. A very slow mineralizable N, with less than 10 percent of the N available in the first growing season (Cooperband 2000), also limits its use.

Most compost is now used for gardens and landscaping, or for the production of artificial soils designed for growing specialty crops in containers. About 95 percent of the compost produced by Douglas, Georgia, goes for landscape use, lawn establishment, and new site development, with only about 5 percent being used for row crops and horticultural crops.

Composts are finding special niches, especially for contaminated, degraded, and low-pH soils and soils of low fertility. For example, composted biosolids have improved the establishment of turf grasses (Kentucky bluegrass and perennial ryegrass) on disturbed urban soil sites (Loschinkohl and Boehm 2001). Very high rates (about 50 tons/acre) have been used successfully on some problem soils, but care must be taken that the high rates do not introduce excessive salts.

The use of composts on farmland will probably increase due to greater availability. The recent trend to compost urban organic wastes and sewage sludge may make composts more available for use in commercial vegetable, fruit, and ornamental plant production. It appears that if composts with suitable criteria are selected, their use for ornamentals and probably vegetables and fruits will increase. In addition, partially processed compost that has not undergone its full cycle of changes may provide extra weed control, also useful in fruit and vegetable production. Using compost as extra mulch to help suppress weeds for vegetables grown in a no-till operation could provide for a substantially increased market for compost. Greater use of compost can be expected if plastic mulch use is reduced because of pollution problems.

Also, composts would find much greater markets if a uniform, high-quality product could be maintained. A fully mature compost made from uniform materials would help provide a more uniform product with greater nutrient availability and increased disease suppression that should provide for better crop production.

SEWAGE EFFLUENTS AND BIOSOLIDS

About 6 trillion gallons of sewage effluents are produced annually in the United States. The effluents, consisting of both liquids and solids, contain important nutrients and organic matter that could be highly beneficial for soils. Unfortunately, several components of sewage are potential environmental hazards because

1. the C, N, and P can eutrophy water in streams and lakes;
2. nitrates can leach into groundwaters;

3. salts can accumulate or leach into groundwaters;
4. pathogenic organisms can be a source of disease in humans; and
5. heavy metals (Cd, Cu, Ni, Pb, and Zn) can harm soils and foods.

These wastes must be carefully treated and handled to be of value to agriculture.

The disposal of human wastes is regulated by the EPA and individual state agencies. Generally, these organizations require that waste materials be treated to destroy disease-causing organisms and reduce the attraction of flies and other vectors of human disease. Composting or heating sludge or biosolids greatly reduces the pathogens. Drying or digesting the biosolids aerobically or anaerobically tends to solve the vector problem. Alkali additions and storage aid in controlling vectors and is effective against various pathogens.

Sewage effluents are usually treated in one or more of the following ways to provide an acceptable product. Initially, the effluent is treated with lime, iron, or aluminum salts to remove 50-75 percent of the solids. In a secondary treatment, the effluent is aerated by spraying it over rocks or gravel beds, allowing it to trickle slowly, or it is aerated mechanically. The secondary treatment removes 80-95 percent of the solids and 20-50 percent of the inorganic salts. A tertiary treatment that adds a polyelectrolyte or uses activated carbon or electrodialysis removes 98-100 percent of the suspended solids, 95 percent of the phosphates, and about 50 percent of the nitrates. The removed materials are part of the sludge or biosolids.

Liquid and solid components are being used for agricultural purposes, mostly in liquid form or as a partially dried cake or sludge. Current processes remove enough of the water to produce two relatively dry products. Sewage sludge is a high-moisture product of low analysis that might be useful on land very close to the source, but application of large quantities far from the production site is laborious and expensive. In addition, this product carries so much moisture that it is often necessary to wait for some time after it is applied before the soil can be utilized. The other product (activated sewage sludge), which is aerated after being inoculated with microorganisms, has a higher analysis and is much easier to handle. It is used commercially primarily for lawns, landscaping, and growing ornamentals. Typical sludge contents of N, P_2O_5, and K_2O were given in Table 3.4.

Although activated sludge has a much higher analysis, its relatively low nutrient content compared to most commercial fertilizers makes its use unattractive for most crops. Also, the potentially high content of heavy metals in these materials may limit their use to nonfood crops unless the metals can be prevented from entering the waste system or nullified by treatments. Even for nonfood crops, addition of wastes with elevated metal content needs to be limited to soils with pH values near or above 7.0 to limit the harmful effects of several heavy metals on the crops.

The liquids or effluents after primary or secondary treatment are piped and applied by irrigation primarily to tree farms and farms producing animal feeds and forages. Municipalities maintain farms for disposal purposes, underwriting some of the cost of disposal by the agricultural commodities produced on the farm.

Not all soils are suitable for such distribution. Ideal soils are nearly level, have a permeable loam topsoil over a deep open sandy subsoil, and have a C horizon deep enough that the soil is at least 6 ft above the seasonal high-water table. Even with ideal soils, it is desirable to monitor groundwaters by testing shallow wells downslope for nitrates, abandoning or delaying additional applications if nitrates are 10 ppm or more.

The principle concerns with continued use of wastewaters are (1) potential health problems, (2) the possibility of water pollution, and (3) the buildup of toxic levels of heavy metals. Biosolids present similar problems, except that the aerated sludge is practically free of disease organisms. The problem of disease transmission can be reduced by treatment prior to its use. Pollution problems can also be mitigated by choosing soils suitable to handle the effluents and by periodic testing of groundwaters.

Heavy Metals

The presence of the heavy metals Cd, Ni, Cu, Zn, and Pb in effluents and sludge or biosolids can cause problems with plant and animal health. The heavy metal problem is reduced by limiting disposal to the capabilities of the soil (Table 7.6), which are affected by its CEC.

The nonmetallic elements As and Se and the metallic elements Cr, Hg, and Mo may be present in harmful concentrations in effluents

TABLE 7.6. Safe* cumulative amounts of heavy metals from biosolids applied to cropland (lb/acre)

Heavy metal	Loamy sands	Loams	Silt loams	Clay soils	Organic soils
Cadmium (Cd)	0.07	0.14	0.27	0.36	1.3
Nickel (Ni)	0.98	1.96	3.8	5.8	19.2
Copper (Cu)	2.32	4.6	9.4	14.0	46.8
Zinc (Zn)	4.7	9.5	18.9	28.4	94.6
Lead (Pb)	9.5	18.9	37.8	56.7	189.1

Source: "Guidelines for the Application of Sewage Sludge to Land." University of Maryland Agronomy Mimeo # 1976.
*Based on CEC in meq/100 g of soil, assuming 5 for loamy sands, 10 for loams, 20 for silt loams, 30 for clay soils, and 100 for organic soils.

and biosolids. Upper limits of these elements as well as the heavy metals have been set for sludge by the EPA and are given in Table 7.7.

Agricultural Use

Various trials indicate that both effluents and sludge have agricultural value. Sludge or biosolids (as it is now commonly designated) compares well with manure. Possible exceptions include poorer crop response to biosolids due to a lower K level than most manures, and occasionally to excess heavy metals.

The fertilizer value of both effluent and biosolids will vary with the source of material and its treatment. Evaluating biosolids, stabilized with lime, for mature 'Hamlin' oranges indicated that a 10.7 ton application per acre supplied more lime than was needed, and about all the necessary N and P_2O_5, but lacked about 185 lb of K_2O needed for the crop. Using 1999 figures, it was calculated that the 10.7 ton application of biosolids per acre supplied the equivalent of $63 of fertilizer and lime. It was estimated that the total cost of lime and fertilizer needed for the crop was about $100 (Roka et al. 1999).

The fertilizer value of micronutrients contained in biosolids or the long-term physical and biological effects on soil were not calculated in the above study. As all of these are considered, it may be desirable to use biosolids, if costs of transportation and handling can be minimized, but attention must be paid to heavy metals and possible excess

TABLE 7.7. Upper limits of heavy metals for sewage sludge as set by EPA*

Metal	Upper limits in sludge (ppm)	Cumulative loading rate in sludge** (lb/acre/year)	Annual loading in soil*** (lb/acre/year)
Arsenic	75	36.6	1.79
Cadmium	85	34.8	1.78
Chromium	3,000	2,679.0	133.95
Copper	4,300	1,339.5	66.97
Lead	840	267.9	13.39
Mercury	57	15.1	0.75
Molybdenum	75	16.1	0.80
Nickel	420	375.0	18.75
Selenium	100	89.3	4.46
Zinc	7,500	2,500.4	125.02

Source: Adapted from Wallace, A. and G. A. Wallace. 1994. A possible flaw in EPA's 1993 new sludge rule due to heavy metal interactions. *Communications in Soil Science and Plant Analysis* 25(1-2): 129-135. Reprinted by courtesy of Marcel Dekker, Inc.
*Sludge or biosolids cannot be used on land if any one level is exceeded.
**Metal concentration in sludge must be known and when loading rate has been reached for any one metal, use of sludge must be discontinued.
***If any metal exceeds its "clean" concentration limit.

salts. As with manure, heavy use of biosolids will probably be confined to areas close to their production, but unlike manure, it may be limited to nonfood crops. Additional use of these products could be made if a greater proportion of the production and transportation costs were underwritten by the communities that need to dispose of the wastes. Also, separating industrial wastes from human wastes would open up a greater market for the products.

PEATS

Peats form under conditions of impeded drainage or high groundwater levels that reduce O_2 levels and slow the decomposition of organic materials. The accumulation of organic matter, rich in lignin and hemicellulose, increases with increasing submergence and cool temperatures. Five types of peat are recognized in the trade: sphagnum moss peat, hypnum moss peat, reed-sedge peat, peat humus or muck, and other peat.

With the exception of peat humus or muck, all peats provide relatively little N, P_2O_5, or K_2O, and are valued primarily for their soil-conditioning properties. Sphagnum peat has an acid reaction and very high MHC (15-30 times its weight). Although reed-sedge peat has much lower MHC and may not be acidic, it is highly prized for golf courses and landscape use, partly because of its longer-lasting properties. Some uses for the different peats are suggested in Table 7.8.

Because of high costs, most peat use will probably be limited to landscape purposes, golf courses, and potting mixes for producing ornamentals and transplants. In recent years, a great deal of the peat formerly used for potting mixes has been replaced by composted pine bark and some other wastes.

HIGH-ANALYSIS ANIMAL AND OTHER WASTES

Applications of various animal waste or crop by-products can contribute much of the N needs of the succeeding crop. Amounts of N, P_2O_5, and K_2O of some of the more common legumes and animal products are given in Tables 3.3, 3.4, and 3.5. But as with OM and manures, the major release of these nutrients is affected by microorganism activity, and only about one-third to one-half of the N can be expected in the first crop year.

The relatively high N value of some of these materials (animal tankage, cocoa tankage, dried blood, fish meal, hoof and horn meal, shrimp scrap, soybean meal) or the K_2O value of others (cotton hull ashes) limits their application to about 1,000 lb or less per acre. Although valuable as nutrients, the relatively low rate per acre makes them of dubious value for increasing SOM. Also, many of these products have disappeared from the fertilizer market due to their greater economic value as animal feeds.

By-Products of Food Processing

Many by-products are generated in food processing besides the high-analysis materials previously listed. The potential amounts of

TABLE 7.8. Value of different peats for agricultural use at common rates of application

Use of peat humus	Rate of soil mix	Sphagnum	Hypnum	Reed	Peat humus (decomposed)
Soil conditioning	2 in layer*	fair	good	good	good
Topdressing lawns, golf courses	1/8-1/4 in layer*	fair	good	good	good
Surface mulch	2 in layer	excellent	fair	fair	poor
Potting soil mix	50% peat or 50% vermiculite or soil	excellent	good	good	fair
Golf green soil mix	80% sand, 10% clay loam, 10% peat	poor	good	good	excellent
Rooting cuttings	50% peat, 50% vermiculite	excellent	good	good	poor
Seed flat germination	pure mulled	excellent	good	fair	poor
For acid-loving plants	25% mixture in soil	excellent	NR	good**	NR
For acid-intolerant plants	25% mixture in soil	okay if limed	good	good***	good
N source	Soil mixes, topdressing	poor	good	fair	good

Source: Adapted from Lucas, R. E., P. E. Ricke, and R. S. Farnham. n.d. "Peats for Soil Improvement and Soil Mixes." Extension Bulletin No. 516, Farm Science Series, Cooperative Extension Service, Michigan State University, East Lansing, MI.
NR = Not recommended
*Worked into soil
**Good if pH <4.8
***Good if pH >4.8

these by-products are enormous, with estimates of 8,007,000 tons from processing citrus (orange, grapefruit, lemon, tangerine, tangelo, temple, lime), 970,000 tons from large fruits (apple, peach, plum and prune, pear, and apricot), 1,035,000 tons from small fruits (grape, cranberry, sweet and sour cherry), 952,000 tons from nuts (almond, walnut, pecan, macadamia, and hazelnut), 10,506,000 tons from vegetables, and 39,034,000 tons from processing grain seeds (soybean, cottonseed, peanut, flax, and sunflower) for oils. The estimates of by-products from animal slaughter and processing are also huge: 8,147,000 tons from cattle, 3,696,000 tons from hogs, 72,000 tons from sheep and lamb, and 5,666,000 tons from poultry. There are other wastes from sugar, coffee and tea processing, and 842,000 tons from processing fish and shrimp (Barker et al. 2000).

The by-products are solid residuals or wastewater. Both products can vary tremendously depending upon the food item, how it is handled prior to processing, and treatment during the processing, which can vary considerably with different processing plants.

The solid residuals consist of many different materials, such as

1. rinds, pulp, seeds, juices, stems, and residues remaining after extracting juices from citrus processing;
2. fruit skins, pits, pomace, presscake, meals, peelings, cores, seeds, or corn cobs from fruit and vegetable processing;
3. bagasse from sugar processing;
4. hulls and leaves from nut processing;
5. animal feces, urine, blood, fat, meat scraps, bone, hooves, hair, horns, dead animals, paunch contents, feathers, or entrails from animal processing; and
6. skin, scales, fins, fish heads, or shells from fish and seafood processing.

These wastes may contain considerable quantities of plant nutrients and large amounts of organic matter, and probably are a great potential source for replenishing SOM. Some of these solids (primarily from vegetable, slaughterhouse, and shrimp and fish processing) are already being used as fertilizers or composts. Some of the solid by-products are also being used for (1) animal feed (citrus pulp, corn gluten, oilseed meals and presscake, and slaughterhouse wastes), (2) processing plant fuel (sugarcane bagasse), and (3) fiberboard (sugarcane bagasse).

More of these wastes should be moved into agricultural application, but much more research needs to be done as to their effectiveness before large quantities of low-analysis materials can be used economically on large acreage. The problem of moving more low-nutrient-content by-products onto the land is compounded by seasonal availability and/or the overly wet condition of much of the material. Both of these shortcomings probably could be overcome by composting the residuals, but often some dry material would have to be blended in with the by-product to create an acceptable product for land distribution.

Wastewater from food processing also has potential for increasing SOM. The wastewater consists of washings, blanching water, and residuals from the product being processed. Obviously, composition varies with the food being processed and the processing procedure, but most wastewaters have a high biochemical oxygen demand (BOD) in a five-day period and a possible high chemical oxygen demand (COD).

The BOD is a measure of the organic matter that is readily oxidized. It is important because it can affect O_2 levels of the medium used for wastewater disposal, since the application rate of OM can exceed the orderly rate of decomposition by microorganisms. Discharged into bodies of water, large amounts of wastewater with a high BOD can so deplete O_2 levels that fish and other aquatic animals are killed. Discharged onto land, a fast rate of discharge can clog soil pores, resulting in reduced infiltration, and produce anaerobic conditions that can affect crop growth adversely.

The COD is a measure of the O_2 required to oxidize dissolved mineral and organic matter. It can affect the oxidation-reduction level of the soil sufficiently to affect plant growth adversely.

In addition, wastewater, depending on the food being processed and mode of processing, may have a high pH (from lye used for peeling), a low pH (from sauerkraut and pickling processes), high N and P content (primarily from slaughterhouses), excessively high temperatures, various contaminants such as pesticides, fungicides, pathogenic viruses, and bacteria (primarily from meat and poultry processing), excess heavy metals, high Na content (from lye used for peeling or salt used for pickling or sauerkraut processing), and oil and grease (from meat or oilseed processing).

Wastewater is being applied to agricultural land, with much of it going to permanent pastures but increasing amounts to row crops. Some of the problems of high BOD or COD are overcome by removing a great deal of the solids by settling, vacuum filtration, or centrifugation. The removed solids often can be used as livestock feed or applied directly to the land. Characteristics of soils receiving wastewater should be similar to those suggested for effluents from biosolids (see Sewage Effluents and Biosolids). As with sewage effluent disposal on the land, careful monitoring of the soil is needed to see that application rates do not exceed the land's capacity to handle them. Eliminating use of wastewater with excessively high Na or heavy metal concentrations and monitoring wastewater, so that low-pH wastewater could be applied to high-pH soils or high-pH waters to acidic soils, should lengthen the period of safe wastewater irrigation.

The potential for using waste solids and wastewater for improving SOM is enormous. Just how much more can be economically diverted for improving SOM is problematic. Present attitudes and ecological laws are favorable for such use rather than landfills. But practical means of getting them applied on the farm still need to be worked out.

Wood Products

A number of wood products can be used as soil amendments, including pine needles, wood bark, wood chips, sawdust, and wood fiber. All of these materials tend to be acid-forming and are used as mulches for acid-loving plants. Pine needles are especially prized for maintaining azaleas. Wood chips, bark, and sawdust are used for landscaping and as a mulch to help establish small-seeded plants and those that are difficult to establish. Bark and sawdust are also used as a partial substitute for sphagnum peat in loamless mixes for pot culture.

Wood products may induce poor growth in a number of plants due to their content of allelochemicals and/or their wide C:N ratios. The ill effect can be largely overcome by composting. Composting pine bark has been so successful that many of the prepared soil substitutes used for potting mixes contain substantial amounts of composted pine bark.

Composts made of softwood bark tend to be more open than hardwood bark because the softwood is more resistant to bacterial decay. This resistance, along with a natural tendency to resist water absorption, tends to provide excess drainage from cultures filled only with softwood compost, and make it unsuitable for most plants except epiphytes. Mixing softwood compost with varying amounts of sphagnum peat can make it more suitable for a wide variety of plants.

Considerable amounts of wood are used in making paper and paperboard. Much of the by-products from these industries are being used as fuel to power various processes, resulting in destruction of the OM but leaving ash. The ash has potential value for agriculture but no direct value in preserving SOM. On the other hand, waste treatment plant residuals may have some value for SOM, but the low nutrient analysis (Table 7.9) probably will limit its use to nearby locations or the forest floor.

TABLE 7.9. Analyses of waste treatment plant residuals from Kraft pulp and paper mills

Nutrient	Range		Medium	
	%	ppm	%	ppm
N	<0.1-21		3.2	
P	<0.1-15		1.4	
K	0.02-65		0.23	
Ca	0.1-25		2.7	
Mg	0.03-2.5		0.4	
S	0.6-0.15		1.1	
B		ND-310		5
Cu		6.8-3,120		475
Fe		1,000-15,400		1,700
Mn		8.5-13,200		340
Mo		ND-68		5
Zn		38-68,000		873

Source: Based on partial data supplied by Edwards, J. H. and A. V. Someshwar. 2000. Chemical, physical, and biological characteristics of agricultural and forest by-products for land application. In J. M. Bartels and W. A. Dick (Eds.), *Land Application of Agricultural, Industrial, and Municipal By-Products.* Soil Science Society of America Book Series #6. Madison, WI: Soil Science Society of America, Inc. Reproduced by permission of the Soil Science Society of America, Madison, WI.

Chapter 8

Placement of Organic Matter

The manner in which organic matter is added to soil has a strong bearing on its effectiveness. Adding it to the surface where most of it is retained as mulch provides maximum moisture retention from rain and irrigation while reducing water loss from evaporation and contributing greatly to reduction of surface temperature, compaction, and erosion. Mixed with most of the topsoil, as is usually done with moldboard plowing, added OM provides the most rapid release of nutrients while supplying short-term improvement in infiltration and internal structure, but provides little or no erosion protection and leads to the most rapid loss of SOM. Placed in the bottom of furrows or behind chisel plows, OM offers maximum improvement of structure at lower levels, but does nothing for erosion control and little for early nutrient release. Putting OM in furrows tends to be a slow and expensive process.

INCORPORATING OM

Incorporating organic material, as is usually accomplished by moldboard plow, disk, or rototiller, provides for rapid decomposition of the material with quick release of the nutrients. Moldboard plowing and disking also improve soil structure throughout most of the plow layer. Unfortunately, the advantages of quick nutrient release and improvement in soil structure are short-lived. The more rapid destruction of OM leads in time to increased erosion and poorer structure, with all the resultant problems of compaction, poor water infiltration, poor air and water-holding capacities, and poor air exchange.

Deep Placement of OM

Organic materials can be added in furrows opened by moldboard plows or subsoilers. Some benefits can be obtained from the practice if subsoils are compact, since the organic materials can provide for longer periods of open subsoil. But the practice usually slows the tillage process. Unless materials are shredded and amounts limited, the practice can also impose problems of water movement in the soil. It is seldom used because of these limitations.

Surface Placement or Mulch

There is mounting evidence that the best placement for OM is to leave it at the surface as mulch. If present in sufficient quantities, it can substantially reduce erosion, lower soil temperature, control weeds, improve water infiltration and conservation, and act as a habitat for beneficial insects.

PLASTIC MULCH

Mulches may be derived from natural organic materials, such as crop residues or composts, or from processed materials, such as paper or plastic. Plastic mulch is being used on vast acreage of vegetables and some specialty plants, where it has gained ready acceptance because of its advantages, which include (1) reduction in leaching losses, primarily of N but also K, Mg, B, and SO_4-S in some light soils; (2) weed control; (3) more efficient use of fumigants; and (4) reduction of evaporation losses from the soil surface. About 21,000,000 lb of polyethylene mulch were being used in Florida alone during the early 1990s (Servis 1992).

Loss of SOM with Plastic Mulch

Use of methyl bromide fumigation has been highly effective in controlling a number of weeds, nematodes, and several soilborne diseases and insect pests in vegetable and other relatively high-priced crops. The effectiveness of the fumigation is enhanced considerably by use of plastic mulches, which retain the fumigant in the soil long enough for good control and provide other benefits such as moisture retention, reduction of nutrient losses by leaching, and reduction of surface compaction. Unfortunately, the senior author's experience with the continuous use of plastic mulches plus methyl bromide fumigation for a number of years indicates a drop in SOM with resulting problems of soil structure on all but sandy soils. Problems of water movement in the soil, increased compaction, and an increase in certain insect pests and soilborne diseases accompanied the deterioration of structure. Some of these problems abated after a cover crop (sorghum) was introduced between the primary crops on some of the acreage, but much remains to be done to completely correct the situation.

NATURAL ORGANIC MULCHES

Replacement of Plastic Mulch with Organic Mulch

There are other difficulties with continued use of plastic mulch. Disposal of spent mulch is becoming difficult and expensive. Also, plastic mulch farming requires large inputs of costly chemicals. Methyl bromide, the fumigant of choice, is being phased out, and there are questions whether alternate fumigants can do as well. The overall aim of maintaining a suitable SOM level is difficult to meet because reduction of SOM appears to be hastened by increased soil temperatures under most plastic mulches. Also, SOM loss probably is hastened by the extensive tillage prior to laying the plastic that is required to obtain sufficient aeration for dispersion of the fumigant.

Advantages of Natural Organic Mulch

The substitution of natural organic mulch for plastic mulch may resolve some of the difficulties. Although not meeting all the advantages of plastic mulch, it may offer some others that can provide better net returns.

Natural organic mulch can also conserve moisture loss from the soil surface and provide some weed control. It cannot provide the same reduction in nutrient leaching as the plastic mulch. Nor can it provide the short-term reduction of soilborne pests provided by soil-applied fumigants. Nevertheless, there have been numerous reports of lower costs of fungicides and insecticides for the primary crop with increased use of cover crops left as mulch. Although the system does not decrease soil nematodes, they appear to cause little or no apparent damage (Phatak, private communication).

The use of natural organic mulches makes major changes in the upper or Ap soil horizon. Placement of organic materials on the surface, either by landscaping or allowing residues of the primary crop or cover crop to remain on the surface through the use of minimum or no-tillage, makes major changes in the upper soil layer compared to conventional tillage, which incorporates most organic materials.

Improvement of Soil Moisture

A primary effect of all mulches is the improvement of soil moisture in the upper surfaces of the soil, but natural organic mulches usually provide better soil moisture than plastic. The natural mulch reduces evaporation from the surface as does the plastic, but it has the advantage of also reducing storm runoff and increasing infiltration. Natural organic mulches may not be better, however, if moisture is limited and the cover crop used to produce the mulch is allowed to grow too long before it is terminated, thereby depleting soil moisture needed for the primary crop. Moisture may be even more depleted if a living mulch is not terminated and is allowed to grow, competing with the primary crop. Terminating the cover crop sooner or, in the case of living mulch, restricting its growth by cultivation, mowing, or use of herbicides can overcome this handicap.

Natural organic mulches improve moisture to a greater extent than if OM is disked or plowed into the soil. Much of the extra storage of water with organic mulches as compared to incorporated OM appears to be due to reduced losses from evaporation, since basin listing, which largely prevents runoff by forming basins, fails to store as much water as straw applied as mulch. In a study conducted at Lincoln, Nebraska, such water also penetrated deeper into the soil where a mulch was used (Table 8.1).

The extra soil moisture present under organic mulches as compared to plastic mulch is usually an asset, especially when rainfall may be in short supply. It is of particular importance in dryland farming where soil moisture usually is the primary limiting factor for producing a satisfactory crop, and penetration by small amounts of rainfall can affect

TABLE 8.1. Water storage as affected by placement of straw and type of tillage

Treatment		Precipitation stored*		Depth of water penetration (in)
Straw	Tilth	in	%	
+	mulch	9.7	54.3	71
+	disked in	6.9	38.7	59
+	plowed in	6.1	34.1	59
−	disked	3.5	19.6	47
−	plowed	3.7	20.7	47
−	basin listed**	4.9	27.7	59

Source: Adapted from Unger, P.W., G.W. Langdale, and R. I. Papendick. 1988. Role of crop residues—Improving water conservation and use. In *Cropping Strategies for Efficient Use of Water and Nitrogen.* ASA Special Publication # 51. Madison, WI: American Society of Agronomy, Crop Science. Society of America, Soil Science Society of America. Reproduced by permission of the American Society of Agronomy, Crop Science Society of America, and Soil Science Society of America, Madison, WI.
*A total of 17.9 in of water fell during the trial. Percent stored = percent of total rainfall.
**Diked or dammed furrows.

yield. Often, small amounts of precipitation have marginal effects on crop production in semiarid areas, but a study conducted on two soils (Pullman clay loam and a Randall clay) showed that straw mulch has the potential for increasing water storage even from very small amounts of rainfall (Shangning and Unger 2001).

Insulation

Organic mulch acts as an effective insulator. Limiting soil temperatures, especially maximum day temperature, is common with organic mulches but may be absent with plastic mulches. In fact, some plastic mulches (black, red, brown, green, or clear), actually help raise the upper soil temperature. Change in maximum temperatures by use of mulches varies a great deal depending on solar radiation, type and color of plastic or the type of organic mulch, soil type, soil depth, and moisture. In the Beltsville, Maryland area, the average temperature under black polyethylene mulch at a 2-inch depth (5 cm) was 10.3°F (5.7°C) higher and at 6-inch depth (15 cm) was 6.1°F (3.4°C) higher than that at similar depths under hairy vetch residue (Teasdale and Abdul-Baki 1995).

Natural organic mulches are effective insulators, shielding the upper roots from excessively low as well as high temperatures. The insulating effect can save perennial plants during the winter months by preventing temperatures from dropping to dangerous levels. The reduction of soil temperature is of great benefit for many plants during the heat of the summer but may be a detriment in early spring or in the fall, when warming bare soils can extend the growing season.

The reduction of temperature plays an important part in SOM conservation. This lowering of temperature plus the lessened aeration at deep layers by limiting the use of certain implements tends to slow the depletion of SOM.

Improvement of Soil Structure

The large mass of OM and the more stable soil temperature and moisture beneath organic mulches encourages increases in earthworm and microbial populations. The earthworm populations tend to increase channels in the soil, with resulting improvement in water infiltration. The increase in microbial populations, especially those of fungi, greatly improve stable aggregates, increasing water and air storage and movement. Much of the success of reduced or no-tillage relates to the better use of organic residues, keeping much of the residues as mulches for long periods.

Natural organic mulches increase SOM, probably due to a combination of smaller losses from oxidation as upper surface soil temperatures are reduced, limited contact between OM and microorganisms that decompose the OM, and reduced losses from erosion. The increase in SOM tends to support a biological population capable of further improving soil structure. This better structure not only allows for improved water storage, but air storage and air exchange also increase.

The improvement in structure has a profound effect on the utilization of water, air, and nutrients. Mulches, by intercepting large drops of rain or irrigation, reduce soil capping or compaction close to the surface. Organic mulches provide an added benefit as the decomposition of the OM releases substances capable of aggregating soil particles into larger peds, allowing better water and air intake, and their movement in the soil. This utilization is aided by biological activities favored by the mulch, which also helps control insects, soilborne diseases, and damage from nematodes.

Lower Nutrient Requirements

Although plastic mulches allow less leaching of applied fertilizer than organic mulches, natural organic mulches have an advantage in that they can contribute nutrients for the primary crop. The contribution from legumes is greater because of the fixed N, but all natural organic mulches have nutrients, much of which are in organic forms that are released slowly. Comparing the N requirements of fresh-market tomatoes grown on black polyethylene to those grown on hairy vetch mulch indicated that the tomatoes grown on the black plastic mulch required 170 lb/acre (190 kg·ha^{-1}) of N to reach maximum yields versus only 79 lb/acre (89 kg·ha^{-1}) when grown on killed hairy vetch mulch (Abdul-Baki et al. 1997).

Comparative Value of Natural Organic and Plastic Mulches

Despite some disadvantages when compared with organic mulch, plastic mulch combined with drip irrigation perhaps is still the more productive method for growing vegetables and some other specialty crops on most soils, but it may not be the most profitable method. Alternate methods using cover crops, mulches, and reduced tillage are becoming more popular as growers become more familiar with methods, select suitable cover crops for their area, and learn to handle them and the primary crop to obtain maximum benefits. Plastic mulch and drip irrigation may produce better yields for some time to come, but in all likelihood, increasing number of growers will have greater financial returns using natural organic mulches.

TYPES OF NATURAL ORGANIC MULCH

There are two distinct types of organic mulches. Living mulches maintain a cover crop for all or part of growing season of the primary crop. Killed mulches utilize terminated cover crops and other sources of OM, such as crop residues, composts, bark, tree clippings, sawdust, etc., maintaining them for the life of the primary crop. The primary crop can be planted through the mulch, or the mulch can be added after the primary crop is started.

Living Mulches

A number of cover crops can provide plant materials that serve as mulch while the primary crop is growing. Various parts of the cover crop (old leaves, stem sheaths, spent flowers, seeds, seed pods, and plant clippings, produced as the cover crop ages) are mowed or trimmed to keep them from competing with the primary crop, thereby forming a mulch that is renewed for long periods and adds OM.

Maintaining the cover crop for all or a good part of the primary crop's life extends the benefits of the mulch. In addition to the benefits already cited for improved SOM, living mulch also crowds out weeds, provides excellent roadbeds for traffic, and increases the number of beneficial insects. Unfortunately, living mulch can at times compete with the primary crop as it may excessively shade it or remove water or nutrients necessary to sustain the primary crop.

Although beneficial for many crops, living mulches are not suitable in certain situations. As a rule, they are not recommended for short crops, especially those that are shallow-rooted and susceptible to drought. Because they can compete with the primary crop for nutrients and water, they probably would not be helpful for crops grown on sandy or infertile soils.

Living mulches have been used in orchards and vineyards for some time, but their use for vegetable and other row crops is relatively recent and perhaps still open to considerable improvement.

Living Mulches for Orchards and Vineyards

Various groundcovers are grown in orchards and vineyards to provide a living mulch. Crops of hairy vetch, common vetch, crimson clover, red clover, rye, buckwheat, chewings fescue, tall fescue, timothy, Kentucky bluegrass, and ryegrass are commonly used in apple and peach orchards, while crops of Pensacola bahia, hairy indigo, perennial peanut, common bermudagrass, and centipedegrass are more suited for the fruit crops (pecans and citrus) grown in warmer climates. The annual crops usually are used when the tree or vine crop is young, while the perennials can serve as continuous groundcovers as the primary orchard trees age.

The best floor management for pecans appears to be a grass alley maintained between tree rows. The area by the tree rows out to the dripline is maintained free of vegetation by means of both preemergence and postemergence herbicides.

Permanent sods are used extensively for bedded citrus in Florida, primarily to stabilize beds and limit erosion. Perennial peanut, a legume, has been suggested as a viable alternate because of the N supplied as well as its longevity (Futch 2001).

Living mulches used for vineyards in California vary depending upon the need. Grasses, which form a fine network of fibrous roots in the top 18 inches, are favored for erosion control or soil conditioning. Annual cereal grasses, such as barley, oats, rye, or wheat, planted in the fall and making good growth during the winter months, are excellent choices if the vineyard is to be cultivated. Perennial grasses are preferred if the floor is to be left undisturbed. Some of these grasses, such as perennial ryegrass or creeping red fescue, are active during the summer months and can compete with grapevines for water and nutrients. Some grasses, such as Zorro fescue and Blando brome, will reseed themselves if left undisturbed. A number of native perennial grasses, such as pine bluegrass, Idaho fescue, and the needlegrasses, go dormant during the summer months and so offer little or no competition with the grapevines. Legumes are the preferred plants if adding N is the primary concern, but because they have a taproot rather than fibrous roots, they are less desirable for erosion control. Vetches (cahaba white, common, lana, and purple) produce more biomass and fix more N than clovers or medics (bur clover), but a mixture of vetches with a grass using an 80 vetch, 20 grass blend will produce even more biomass. A blend of vetches, bell beans, and peas also produces large amounts of biomass and is popular because it is easy to grow. Unfortunately, the vetches, bell beans, and peas are annuals and will have to be replanted each year, but some legumes, such as rose clover, subterranean clover, and the medics, act as perennials if they are left undisturbed so that they can reseed themselves (Costello 1999).

Competition of mulch crop with the primary crop. Despite improvements in SOM, and water infiltration and utilization, living mulches may lower yields of orchards and vineyards. Some of this reduction appears to be due to reduced root densities of the trees or vines under the living mulch, which probably are the result of competition for air, water, and nutrients as the primary crop ages. In young orchards or vineyards, competition from the groundcover can be reduced or eliminated by maintaining a vegetation-free area around the individual trees or vines. The area needs to increase as the tree or vine ages, but it may be difficult to leave significant living mulch areas that are not restrictive when some plants (pecans, apples, citrus) reach maturity.

The restriction may occur even when vines or trees are very young. In a study conducted in Indiana, first-year grapevines failed to make as much growth with living covers of winter wheat, oats, or hairy vetch as compared to weed-free check (Bordelon and Weller 1995). A somewhat similar result occurred when apples grown with canola planted in mid-August each year and tilled under the following May failed in the first three years to make as much growth as apples grown in areas kept weed free with annual applications of glyphosate (Merwin et al. 1995). Weed control evaluation of several methods for 'Navaho' blackberry also indicated that best early blackberry growth was obtained with hoeing or use of the herbicides simazine or oryzalin rather than living mulches of alfalfa or rye (Whitworth 1995).

A common approach to avoid such competition in vineyards and orchards is to eliminate the groundcover and maintain a weed-free floor using herbicides. This approach, while useful for many soils, has failed under some conditions as lack of cover may lead to compaction or erosion. Compaction can interfere with air and water infiltration, while erosion can lead to loss of topsoil and a decline in yields. A weed-free floor, by limiting the amount of OM added to the soil, may in time lessen SOM sufficiently to adversely affect yields because of soil structure deterioration and redweed water infiltration.

The answer may be better selection of soils and proper terracing before planting long-term orchards or vineyards, although some combinations of chiseling, soil trenching, and limited sod cover with mulches from vines or trees have maintained satisfactory yields.

Restrictions of growth by groundcovers may in some cases be beneficial. A study of groundcovers under peach trees revealed increased trunk cross-sectional area, canopy diameter, total yield per tree, large fruit yield per tree, and pruning weight per tree with increased sizes of different sized vegetation-free areas (3.9-14.0 ft^2) around trees.

Yield efficiency stabilized as trees grew older, with smaller vegetation-free areas producing equally efficient trees (Glenn et al. 1996a). Establishing a full sod after three year's maintenance with a 26.9 ft^2 vegetation-free area appeared to be appropriate for small efficient trees suitable for high-density systems (Glenn et al. 1996b).

Living Mulches for Row Crops

In both orchards and row crops, undue competition from the cover crop usually should be avoided. There are several methods of limiting competition from the living mulch: (1) it may be possible to select a cover crop or time its growth so that it is less competitive with the primary crop; (2) competition from the cover crop can be reduced by limiting its growth through the use of more frequent mowing, some tillage, or low doses of herbicide; or (3) it may be possible to find the cause of competition and supply the limiting factor so that the primary crop will not be restricted.

Recently there have been efforts to evolve systems of living mulches for various row crops. The introduction of certain herbicides capable of markedly slowing the growth of the living mulch without killing it, the selection of cover crops offering little or no competition with the primary crop, and the development of planters capable of planting through trash have spurred these efforts.

A novel use of herbicides in Hawaii is an example of a practical program of maintaining a living mulch with little competition with the primary vegetable crop. Small widths (8-12 in) of established rhodes grass, corresponding to future planted rows of vegetables, are killed with herbicide. The killed strip is planted to such vegetables as peppers or cucumbers. The grass sod between vegetable rows is maintained but suppressed with low concentrations of Poast herbicide to keep it from becoming too competitive with the vegetable crop. The sod strip keeps weeds from proliferating and provides other benefits such as erosion control, sanctuary for beneficial insects, and an increase in organic matter.

A variation of the living-mulch system uses cool-season crops that die out naturally in the Southeast as the season advances. Crimson clover planted late in the fall will get well established as the fall crop is maturing and make substantial growth in winter and early spring before it dies out. Crops planted in the dead strips will not have any serious weed competition until the crop is well established. Although dead, the crimson clover left in place will still provide weed control as well as many other benefits.

As with the killed mulch control, special tools are needed to prepare suitable strips for planting using coulters to cut through the dead mulch, plant the seeds, and firm the soil suitably so that germination is normal.

Selection of cover crops. Cover crops suitable for orchards are not always desirable for vegetables, and some cover crops suitable for certain vegetables may not be satisfactory for others. Generally, the cover crop should be selected for the primary crop. Vining cover crops, such as hairy vetch, may be suitable for tall crops, such as corn, but shorter vegetables would be better served if shorter cover crops, such as red, white, or subterranean clovers, were used.

There are other criteria for selecting the cover crop, some of which apply to vineyards and orchards as well as row crops. Generally, the cover crop should be able to germinate and tolerate some shade. It should not be subject to the same pests as the primary crop, nor should it interfere with the harvest of the primary crop. Cover crops for certain vegetables, especially those that may be harvested over a period of time, need to stand up to traffic. For example, soybeans, although an efficient cover crop, are not suitable for multiple harvested crops.

Cultural methods. Different cultural methods are used for handling the mulch and primary crop, depending on the type of crops, climate, and soil type. Commonly, the cover crop is sown before the primary crop. The cover crop is strip tilled, and the primary crop is planted after a short period in the tilled strips. In other cases, the cover is planted at the same time as the primary crop or after the primary crop is established. The mulch crop is usually interseeded in the aisles at time of planting or later, after the primary crop is established.

Planting the cover crop before the primary crop is started requires preparing narrow strips of the cover crop for planting by means of tillage, mowing, or application of herbicides. Herbicides are more apt to provide satisfactory planting areas in the shortest time. Tillage can provide satisfactory planting areas but may be less satisfactory with some cover crops, particularly if they are mature. Mowing is the least satisfactory for providing suitable planting areas. The primary crop is planted in the cleared strip, keeping the remaining cover crop but controlling it by tillage, mowing, or low doses of herbicide.

Cover crops planted with or after the primary crop will make more rapid development if they are drilled rather than broadcast. Those planted after the primary crop are usually seeded after the primary crop is allowed to grow for several weeks or at the last cultivation in order to minimize competition between the crops.

Suitable cover crops for living mulches in row crops. White clover appears to work well as living mulch for sweet corn and tomatoes. It and subterranean clovers offer promise with many vegetables grown in relatively cool climates because of their low compact growth, ability to hold up under traffic, and their fixation of N. Perennial and annual ryegrass are being used with tomatoes, beans, and squash. The perennial seed is more expensive but is more cold hardy. The annual type is preferred if the field is to be replanted to vegetables the following year, but the perennial is better suited if the following crop is to be hay or pasture (Peet 2001). Barley has been used for onions in northern climates, while sod crops such as St. Augustine grass, perennial peanut, and rhodes grass have been used in warmer climates for several vegetables.

Selecting or timing the cover crop to reduce competition. Selecting the proper cover crop and timing its growth so that it will supply ample OM and provide a satisfactory living mulch while limiting its competitiveness is a real challenge. Using local wisdom and building on others' experiences are usually helpful. Selection of cover crops from the handbook *Managing Cover Crops Profitably* (Clark 1998) can aid in selecting desirable crops.

For row crops, a winter cover crop, such as red clover in the southeast, which will grow slowly and eventually die as temperatures rise and the primary crop is stimulated, can be very helpful. By interplanting 24-inch strips of berseem clover or hairy vetch in furrows between wheat in double rows about 8 inches apart on 30-inch beds, wheat production was more than doubled in low-N soils under irrigated conditions in northwestern Mexico. In other studies, common and hairy vetch, crimson clover, and New Zealand and ladino white clover as well as berseem clover interseeded with wheat or barley in low-N soils provided multiple benefits on heavy clay soils while maintaining grain yields (Reynolds et al. 1994).

Limiting the growth of the cover crop. Common methods of controlling the cover crop consist of periodic mowing, tillage, or use of herbicides to limit height and reduce growth. Mowing can limit competition and often is a means of increasing the OM from several crops. Berseem clover cut and left in the field may restrict growth excessively unless the crop is cut with a sickle bar or flailed. It needs to be periodically fluffed with a tedder to stimulate regrowth from lower stems. Buckwheat will make fast regrowth if cut before 25 percent of blooms open. Medic or bur clover will tolerate regular mowing to a height of 3-5 inches. Medium red clover does well with several cuttings in its second year of growth. Mammoth red clover, although it produces a larger biomass at first cutting, does not lend itself as well as the medium red to multiple cuttings. Sorghum-sudangrass hybrids will make adequate regrowth if a 6-inch stubble is left (Clark, 1998).

Light tillage with disks, cultivators, or harrows often can be used to suppress the cover crop. Some of the cover crop can be killed by the tillage, but enough usually is left to regenerate the cover crop.

These discussed methods do not always limit competition. A rye cover crop maintained as a living mulch between rows of pepper grown on black plastic provided good weed control, but reduced yields of peppers as much as 50 percent when compared with clean-cultivated or herbicide-treated middles. Mowing the rye failed to overcome the negative effect of the living mulch (Reiners and Wickerhauser 1995).

In some cases, careful application of herbicide either at reduced rates to curb the mulch or at full rates to kill it is necessary. Reduced rates of Poast are useful for controlling several grasses, and reduced rates of Roundup can be used for both grasses and dicotyledonous plants. Sufficient shielding of the primary crop from the herbicide is necessary to avoid damage to it, unless it is well established that the primary crop can tolerate the rate used.

It has been necessary to prematurely kill cover crops of barley or rye interseeded in carrots or onions grown on organic (muck) soils in Michigan. The cover crops are used primarily to reduce wind erosion, which can at times remove 1-2 inches (2.5-5.0 cm) of soil in one hour. Interseeding the cover crops in carrots or onions is effective in reducing erosion, but onion yields can be reduced if barley exceeds 8 inches or rye exceeds 7 inches in height. Carrot yields are reduced if barley exceeds 16 inches. Two graminicides (Poast or Fusilade) capable of killing grasses and small grains with little or no harm to many dicotyledonous plants are routinely being applied when the small grain cover crop is about 4-5 inches tall, but the cover crop may not collapse for another week. Despite the young age of the cover crop, it is sufficient to effectively control wind erosion at the most critical time of the primary crop's growth (Zandstra and Warncke 1993). A somewhat similar optimum height of 18 cm (7.2 inches) for controlling barley used as living mulch was also obtained with the herbicide (fluazifop-P) for onions grown in North Dakota (Greenland 2000).

Supplying items in short supply. The ill effects of competition between cover and primary crop can in some cases be alleviated by supplying the items that become limiting. Besides light, the factors usually causing reduced growth of the primary crop are insufficient (1) water or (2) nutrients.

1. A common cause of poor growth of the primary crop from competition of the cover crop is the lack of moisture. Roots from the living mulch may extend into the area of the primary crop and remove enough moisture to limit the growth of the primary crop. In fact, the cover crop may take out so much moisture that it is difficult to start the primary crop. As cited above, mowing the cover crop and early tilling a 6-10 inch band for planting the primary crop may be enough to establish the primary crop. Irrigating to supply extra moisture can be helpful in establishing the primary crop and in later stages stimulate its development.

Two aids can be used to ensure maximum benefits from irrigation. Use of soil moisture sensors to determine if moisture is limiting can help make the decision for supplying the irrigation. Drip irrigation placed close to the primary crop will ensure that most of the water will stimulate the primary crop without unduly furthering the competition.

2. The cover crop can remove enough nutrients from the primary crop to curtail its growth. Soil and leaf analyses of the primary crop can pinpoint elements that are in short supply, and these can be added. Adding them through drip irrigation or as foliar sprays directed to the primary crop ensures rapid delivery with little or no benefit to the competing cover crop.

Killed Mulch

Unlike the living mulch, killed or dead mulch uses plant residues, killed cover crops, or other sources of organic matter, such as compost, lawn clippings, shredded yard trimmings, tree limbs, bark, sawdust, leaves, straw, pine needles, and other wastes. Commercial growers most commonly use plant residues or killed cover crops because of savings in costs of handling and transporting materials, although there has been some use of composts because of recent availability. Landscape gardeners and homeowners make use of wood wastes and composts. The wood wastes are favored for many ornamental woody plants because of their appearance and usual acid reaction, which is favorable for some ornamental trees and shrubs. Composts, lawn clippings, and old newspapers are used by homeowners as a killed mulch for vegetable gardens.

Plant residues or killed cover crops have become popular with growers of row crops, especially vegetable growers, and are increasingly popular with cotton and peanut growers. Previous crop residues may be combined with a cover crop that does well in the period between crops. The cover crop is often started by interseeding it in the previous primary annual crop, usually in late summer or early fall, allowing the cover crop to get a good start before cold weather slows growth. The cover crop is terminated shortly before planting the new primary crop by undercutting or rolling the cover crop with a rolling stalk chopper, or by use of a nonselective herbicide, e.g., Roundup or Paraquat. The primary crop is planted with a minimum-tillage planter that can cut through the mulch formed by the killed cover crop and any plant residues remaining from the previous primary crop. The planter usually is equipped with spring rollers capable of firming the soil around the seed or transplant so that it will get off to a rapid start.

A variation of planting in the killed mulch is used to gain a few days' extra growth of the cover crop. Corn, or another seeded primary crop, is no-till planted in the cover crop while it is still standing, and the cover crop is flail chopped several days later, just before the seeded primary crop emerges. The practice gains some extra growth of the cover crop, and in case of legumes, such as hairy vetch, considerable extra N.

Killed cover crop mulches are particularly suitable for no-till vegetables and are gradually displacing polyethylene. There are several advantages in using the killed cover crop mulch, but for many growers, the primary appeal may well be reduced costs of growing the primary crop. While the benefits of N (from legumes) and a more orderly release of nutrients appeal to many growers, the cost of providing an organic mulch from a cover crop such as hairy vetch is only a small fraction of that required for installing and removing a polyethylene mulch. Added benefits of the natural organic mulch, at least for certain situations, are the elimination of fumigants and a greatly reduced need for insecticides and fungicides.

Suitable Cover Crops

Several cover crops are suitable as killed mulches, but the ideal cover crop or mixture of crops no doubt varies with different climates, soils, and primary crops. The length of the growing season can influence which cover crops can be used and the age at which they need to be terminated. Hairy vetch is a leading contender in the eastern United States. It can be mixed with rye or other tall cereals and crimson clover, although a mixture of vetch and rye did not do as well as vetch alone in a comparison of mulches for tomatoes conducted at Beltsville, Maryland (Abdul-Baki et al. 1997). Rye, by itself or mixed with vetch or crimson clover for extra N, is also popular with vegetable growers and is being used by growers of soybeans and several vegetables.

Generally, maximizing biomass production is desired not only because of its effect on SOM but also its importance in suppressing weeds. To suppress weeds, which can be a serious problem with organic mulches, it is necessary to obtain considerable biomass that will persist for a good part of the primary crop life. For example, Abdul-Baki et al. (1997) recommend including 40 lb of rye seed with 40 lb of vetch per acre for greater biomass and longer acting mulch and weed control. Adding 10-12 lb/acre of crimson clover aids weed suppression as well as adding extra N.

Crop Residues As Organic Mulch

A good part of the success of no-till or reduced tillage is due to the minimal soil preparation, which allows some if not all of the residues to remain at the surface. The organic mulch may be derived from cover crops, but for economic or other reasons, mulches for field crops often are derived only from residues of the previous primary crop.

Maximizing the Mulch Effect

Regardless of whether the source is a cover crop or the residues of a previous crop, many farmers would be well served if OM were handled to provide maximum coverage of the soil as a mulch. The proportion of surface covered as the crop is harvested is affected by the crop, its yield, and whether it is grown conventionally or with reduced tillage, but its durability is affected by the crop, the way it was harvested, and the manner in which the soil is prepared.

Extent of coverage. Much of the value of mulches is based on their erosion control, which is closely related to the amount of coverage provided by the mulch. Protection of the soil surface from wind and rain is largely dependent on the extent of coverage by the residue of the harvested primary or cover crop, which is affected by (1) the kind of crop, (2) crop yield, (3) whether the crops were grown under conventional or no-tillage systems, and (4) the tillage systems used for incorporating the crops. As was seen in Table 6.4, the residue from prior crops of corn and soybeans grown in Illinois was closely dependent on the yields of these crops. Furthermore, the mulch tended to build up under continuous no-till systems, adding about 7 percent surface cover for nonfragile crops such as corn and about 3 percent for the fragile soybean crop.

Benefits of different residues. There are often marked differences in benefits of different residues. For example, wheat straw is better for preserving infiltration than German millet, which in turn is better than sorghum residues. Some of the benefits may relate to differences in weight produced by various crops, but often relate to other factors, such as different densities and diameters of the residue and its ability to last over longer periods.

Fragile versus nonfragile residues. The benefits of mulched organic residues increase as their time of persistence increases. The residues that decompose quickly (fragile) will not provide the same benefits as those that persist over longer periods (nonfragile). The fragility of the residue is affected by the manner in which the crop was harvested as well as by the nature of the crop. Some fragile and nonfragile crops are listed in Table 2.2.

The amount of OM left on the surface to form mulch is closely related to the manner in which the residues are handled prior to replanting the primary crop. Maximum amounts are left on the surface if the residues or the cover crop are not incorporated into the soil. Special equipment needs to be used to plant through the trash, which is briefly covered in Chapter 10.

Although more growers are beginning to use no or minimum tillage to grow crops, most crops are still grown conventionally, incorporating residues or cover crops prior to planting the primary crop. The effects of different tillage systems on residue cover vary with the type of crop, tillage implement, the depth at which it is set, and the extent of secondary tillage used to prepare the seedbed.

Nonfragile crops generally provided better coverage than fragile crops, but the extent of coverage of both crop types were influenced by tillage tools. Of the primary tillage tools examined, chisel plows and V-rippers gave about the best coverage, followed closely by mulch tillers with coulters. Disks generally gave poor coverage, although disks with 11-inch spacing set at a 3-inch depth provided fairly good coverage with nonfragile crops. Of the secondary tillage tools, there was little difference in extent of coverage using field cultivators, roller harrows, and mulch finishers, but all three were superior to disks set at 4 inches (Table 8.2).

DELETERIOUS EFFECTS OF ORGANIC MULCHES

The beneficial effects of organic mulches far outweigh the adverse effects. Nevertheless, it is wise to know the possible harmful effects, because many of them can be avoided or lessened by appropriate measures. Most of the adverse effects of organic mulches are due to (1) the high level of moisture maintained in the upper part of the soil, (2) reduced warming of soils, (3) the extension of harmful effects of certain pesticides and allelopathic substances, and (4) increased damage from certain insects, disease organisms, and small animals. Some of these adverse effects, which may be worse with large amounts of organic materials no matter where they are placed, are covered in Chapter 2. But placement as a mulch increases some of the harmful effects and is covered more fully herein.

Limiting Ill Effects of Excess Moisture

The increased moisture at the surface induced by organic mulches may physically delay the planting of crops. The lower soil tempera-

TABLE 8.2. The effect of tillage tools on the proportion of residue remaining of fragile and nonfragile crops having variable original soil cover

Primary tillage tools	Original cover														
	100%		90%		80%		70%		60%		50%		40%	
	F	NF	F	NF	F	NF	F	NF	F	NF	F	NF	F	NF
	% remaining													
Chisel plow w/ 12" space w/														
16" med. crown sweep	55	80	50	72	45	65	38	55	33	50	27	40	22	32
2 × 14" chisel point	50	70	45	65	40	56	35	50	30	42	25	35	20	28
4 × 14-1/2' shovel	45	60	40	55	35	50	30	42	27	35	22	30	18	25
3 × 24' concave twisted shovel	45	60	40	55	35	50	30	42	27	35	22	30	18	25
Mulch tiller w/ coulters 15" spacing w/														
18" low crown sweep	50	75	45	70	40	60	35	55	30	50	25	40	20	32
2 × 14" chisel point	50	75	40	55	35	50	30	45	27	40	22	32	18	25
3 × 24' concave twisted shovel	40	55	35	50	30	45	28	38	25	33	20	27	15	22
4 × 24' concave twisted shovel	35	50	30	45	28	40	25	35	20	30	17	25	14	20
Mulch tiller w/ disk gangs 15" spacing w/														
18" low crown sweep	45	62	40	55	35	50	30	43	27	37	22	30	18	25
2 × 14" chisel points	40	55	35	50	30	45	28	38	25	33	20	27	15	22
3 × 24' concave twisted shovel	35	50	30	45	28	40	25	35	20	30	17	25	14	20
4 × 24' concave twisted shovel	30	45	27	40	25	35	21	30	18	27	15	22	12	18

	Original cover												
	100%		90%		80%		70%		60%		50%		40%	
	F	NF	F	NF	F	NF	F	NF	F	NF	F	NF	F	NF
	% remaining													
Primary tillage tools														
V-Ripper														
30" spacing	55	70	50	65	45	60	38	50	33	42	27	35	22	28
20" spacing	50	60	45	55	40	50	35	40	30	35	25	30	20	25
Disks w/ 11" spacing														
3" deep	35	70	30	65	28	60	25	53	20	45	17	37	14	30
6" deep	20	40	18	35	16	30	14	28	12	25	10	20	8	15
Disks w/ 9" spacing														
3" deep	30	65	27	60	25	55	21	50	18	42	15	35	12	28
6" deep	18	35	16	30	14	28	12	25	10	20	9	17	7	14

TABLE 8.2 (continued)

Secondary tillage tools	\multicolumn{2}{c	}{75%}	\multicolumn{2}{c	}{60%}	\multicolumn{2}{c	}{50%}	\multicolumn{2}{c	}{45%}	\multicolumn{2}{c	}{40%}	\multicolumn{2}{c	}{35%}	\multicolumn{2}{c}{25%}
	F	NF	F	NF	F	NF	F	NF	F	NF	F	NF	F	NF
	\multicolumn{14}{c}{% remaining}
Field cultivator														
4.5" spacing w/ 4' S-Tine sweep	*	*	*	*	*	*	*	*	22	30	20	28	13	20
6" spacing**	*	*	*	*	30	42	27	38	24	34	21	30	15	21
9" spacing***	*	*	40	55	32	45	30	40	25	36	22	31	16	22
Roller harrow w/														
2-1/2" sweep	*	*	*	*	22	35	20	30	18	28	15	24	11	17
1-3/8" × 7-1/2" reversible shovel	*	*	*	*	17	30	15	27	14	24	12	21	8	15
Mulch finisher														
8" spacing/9" sweeps 4" deep	40	45	33	35	27	30	25	27	22	24	20	21	13	15
Disk 4" deep w/														
7-1/4" spacing	16	37	15	30	12	25	11	22	10	20	8	17	6	12
9" spacing	22	40	18	33	15	27	13	25	12	22	10	20	7	13

Source: John Deere Tillage Tool: Residue Management Guide C. 1991. Addison, IL: Datalizer Slide Charts, Inc.
F = fragile crops; NF = nonfragile
*Residue levels too high for equipment to function properly
**With 7" or 9" sweeps
***With 10" or 12" sweeps

tures induced by the insulating effects of the mulch and prolonged by high moisture can delay emergence of the crop. Such delays can substantially lower the yields of some crops, such as corn, when grown in cold regions with a limited growing season. Delays also can be detrimental in other areas, where early planting is required to produce a crop in time to command a premium price. Moving a small part of the mulch from a strip 6-10 inches wide from the intended seeding or planting row several days ahead of seeding or transplanting will often allow sufficient drying of the soil to permit earlier planting. The increased warming of the soil by removing a portion of the mulch also hastens germination and early growth.

Lower Soil Temperatures

Lower soil temperatures resulting from organic mulches can be beneficial in the summer but may be detrimental during cooler seasons, when heating the surface of unmulched soil during the day can warm it sufficiently to extend the growing season or reduce cold damage during the night. Reducing soil temperatures can be detrimental for early vegetable crops, since early crops often have a price advantage. Waiting for the soil to warm enough to promote fast growth often can result in missing the best price. An extended period of cool soil can make corn growing in many northern parts of the United States a risky business, as the season may not be long enough to ripen the corn.

Night damage as temperatures dip near the freezing point is primarily on plant shoots and leaves as they, unlike the roots, do not benefit from temperature modifications of the mulch. On the contrary, they are much warmer during the day because of reflected solar radiation, but cooler during the night because of heat radiation from plant surfaces to the cold air.

Low soil temperatures due to mulch can be mitigated for some plants by removing a portion of the mulch close to the plant. Removing the mulch during the early part of sunny days to allow maximum warming during the day and replacing it at nightfall can save sensitive plants if temperatures should drop too low. Although it is not commercially practical, homeowners may want to use this approach to save valuable plants.

Commercial growers can lengthen the growing season somewhat even with mulches by using zone tillage, which prepares a small part of the row for early planting while leaving the remainder of the row under mulch. A lesser alternative that moves aside a part of the mulch at planting time may be sufficient to extend the growing season for borderline areas.

Allelochemicals and Pesticides

When organic matter remains on the surface, the decomposition of organic materials is slowed, thereby causing some harmful aspects of organic matter to continue longer.

Plants contain many different allelopathic substances that selectively inhibit growth of certain soil organisms, limit germination of many different seeds, and restrict growth of many successive plants. Water extracts of composted waste materials yield substances that inhibit germination and early growth of seedlings and cause darkening and death of root cells (Patrick and Koch 1958).

Placing organic materials as mulch can extend the effective life of some allelopathic substances, but because they are not in intimate contact with plant roots, they may cause fewer problems than if the OM was worked into the soil and the crop was planted soon afterward. Moving the organic mulch so that seeds or transplants can be placed in a zone relatively free of allelochemicals often can eliminate much of the potential harm.

The slower breakdown of mulch allows certain pesticides to persist for longer periods, thereby possibly affecting the germination and early growth of sensitive crops. Relatively little work has been done on the subject, but it appears as if the problem may be more serious if an herbicide has been applied recently and little time has elapsed for its normal degradation. The recent emphasis on proper disposal of yard waste has prompted studies with mulches of grass clippings previously treated with herbicides. The results indicate that grass clippings, treated with 2,4-D, dicamba, and MCPP a short time before mowing are harmful to growth and development of tomato, cucumber, salvia, and marigold (Bahe and Peacock 1995). Since activity of these herbicides is lessened with time after application, one would expect considerable reduction of damage by allowing some time between application of the herbicide and use as mulch. Moving the mulch to provide a 6-10 inch mulch-free strip, as suggested previously, would also lessen the problem. A more positive approach probably would be to compost the clippings prior to use if the extra costs of compost can be justified.

Composting Organic Materials Used As Mulch

Composting organic materials before using them as mulch seems to negate the effects of many pesticides and allelopathic substances, while limiting the destructiveness of plant insects and diseases. It is important that the composting process be completed prior to using the compost to eliminate the damage from these sources. Composting pine bark prior to its incorporation into a soil medium eliminates much of the restriction of crop growth that may occur when using mixtures of bark, sand, and peat. Using compost as mulch reduced the incidence of pepper plant loss caused by *Phytophthora capsici* (Roe et al. 1994).

It is important to allow the compost to go through the entire heating and subsequent cooling to eliminate problems of allelopathic substances, insects, plant diseases, and weed seeds and to suppress several soilborne diseases. If in doubt as to the thoroughness of the compost processing, moving the mulch away from the intended seeded or planting area, as was suggested for warming the soil, should largely prevent damage from the allelopathic substances and pesticides that may survive the composting process.

Unfortunately, it is not practical or economical to compost much of the organic matter that will be used as mulch. Allowing some time after the cover crop is terminated or residues are in place before planting the crop also will lessen some of the harmful effects of pesticides, allelochemicals, and pests present on the mulch material.

Pests Associated with Organic Mulches

The presence of insects and disease organisms in OM can also cause problems with the succeeding crop, but as with allelochemicals and pesticides, proper complete composting appears to limit damage from these organisms, giving compost mulch an advantage over mulch formed from residues.

Excess moisture caused by mulch may favor certain diseases of the primary crop. Foot rot of fruit trees, especially citrus, may be favored by the continuously high levels of moisture around the trunk induced by the mulch. Such problems usually can be avoided by leaving an area free of mulch for at least several inches adjacent to the trunk.

Problems Caused by Small Animals

Mulch may also harbor field mice, voles, and slugs. Field mice and voles can injure the bark of fruit trees, but keeping an area free of mulch adjacent to the trunks, as suggested for disease is of some help. Rodenticides may be needed to control large populations. Slugs can damage tender ornamentals and some vegetables, such as cabbage and lettuce. Limiting moisture where possible can reduce the problems, but it is often necessary to start control measures. Pans of beer may be enough to limit serious damage in the home garden, but sprays of metaldehyde or similar materials may be necessary for commercial operations.

PROVIDING PROTECTION FROM PESTS BY USING ORGANIC MULCHES

Despite the fact that some pests are associated with mulch, considerable practical experience indicates that organic mulch combined with reduced tillage, increased cover cropping, limited use of harsh pesticides, and maintaining habitats suitable for beneficial organisms can actually decrease the need for insecticides, fungicides, and nematocides. On a tour of several farms on the coastal plain of Georgia, led by Dr. Sharad C. Phatak of the University of Georgia, the senior author examined crops of pepper and zucchini grown with this system that had excellent insect and disease control with a minimum of one or two sprays as compared to conventional inputs of six or more sprays. Similar results were obtained by growers of peanuts and cotton, with considerable savings in pesticides.

Just what part is played by organic mulch in the reduction of pest pressures is not certain, but organic mulches are known to favor beneficial insects, and the improved diversity of soil organisms favored by the mulch can reduce soilborne diseases and the harmful effects of nematodes.

At least part of this beneficial response appears to be associated with the ability of mulch to produce biologically active soils. Such soils tend to produce crops that resist pest pressures better than crops grown on soils of low fertility, low pH, and poor structure. In this respect, mulch effects are similar to those obtained from incorporated additions of most OM, such as compost, manure, cover crops, and residues, or the presence of large amounts of SOM. There are pluses for mulch, such as better water infiltration, increased moisture levels, and insulation from extremes of temperature.

One of the more important pluses is the ability of mulch to act as a haven for beneficials. In this respect it is better than SOM or most incorporated OM but probably inferior to cover crops. Beneficials are organisms that can favor the primary crop by attacking various pests. Some of the more effective beneficials include the insidious flower bug *(Orius insidiosus)*, minute pirate bug *(Orius tristicolor)*, bigeyed bug *(Geocoris* spp.), spined soldier bug *(Podisus maculiventris)*, convergent lady beetle *(Hippodamia convergens)*, pink spotted lady beetle *(Coleomegilla maculata)*, common green lacewing *(Chrysoperla carnea)*, aphid midge *(Aphidoletes aphidimyza)*, parasitoid wasp (*Encarsia formosa, Diadegma insulare*, and *Eretmocerus eremicus*) and predatory mite (*Galendromus occidentalis, Neoseiulus fallacis*, and *Galendromus pyri*). Several hundred beneficials that specifically attack aphids, beetles, bugs, caterpillars, flies, grasshoppers and crickets, leafhoppers, mealybugs, mites, psyllid scales, thrips, and whiteflies are listed in the *Natural Enemies Handbook* (Flint and Dreistadt 1998), a veritable treasure trove of information about beneficials.

The generalist beneficials, such as insidious flower bugs, ladybeetles, and bigeyed bugs cited in Chapter 6, can exist on many different pests and are not difficult to maintain. When pests such as aphids and thrips are in short supply, they can exist on nectar and pollen of several plants. Other beneficials are very dependent on a particular pest and may be more difficult to maintain if pesticides eliminate all sources of the pest. Maintaining several habitats, such as cover crops and mulches, at all times increases the chances of survival of the various beneficials, ensuring their readiness when pests attack (*Beneficial Bug Guide* 2000).

Another unique benefit of mulch in controlling disease and pests is its ability to markedly reduce the soil splashing induced by rain or overhead irrigation. The protective waxy cuticle layer of many leaves can be breached by soil splashing onto the surface, allowing a ready entry of soil organisms and other organisms into plant tissue and giving both fungal and bacterial diseases a ready start. (More about controlling pests with organic matter is covered in Chapter 11, "Putting It All Together.")

Chapter 9

Conservation Tillage

INCREASING SOIL ORGANIC MATTER

One of the surest ways to increase SOM is to reduce tillage. Reduced tillage, which is also known as conservation tillage, may vary from the almost complete elimination of conventional tillage (CT) to various kinds of reduced tillage (RT). The SOM of the upper 1-2 inches of no-tillage (NT) soils may increase as much as two to three times that of CT soils. Even in the relatively warm climate of Alabama, organic matter in the upper 6 inches of a fine sandy loam soil increased 56 percent with ten years of NT as compared to CT (Edwards et al. 1992). Although varying with different rotations and soils, the increase in SOM with NT in many parts of the Southeast is about 0.1 percent per year until a new equilibrium is reached. Improvement in SOM with NT, although less pronounced with increasing depth, may extend 10-15 inches deep. The depth and amount of improvement with NT over CT depends on the climate, type of soil, amount and kind of residues deposited, and the length of time in NT.

The increase in SOM with reduced tillage appears to be due to a combination of (1) the lower amount of soil O_2 in conservation tillage, as compared to recently plowed or disked soil, which slows OM decomposition by microbial activity and allows more SOM to accumulate; (2) the larger proportion of OM on the soil surface, which reduces the amount of OM in intimate contact with soil microorganisms that can quickly decompose it; and (3) the greater amount of OM on the surface, usually decreases erosion, which is responsible for considerable SOM loss.

Another reason for increases in SOM with NT could be a decrease in organic matter decomposition due to the better stabilization of aggregates as tillage is decreased. Aggregates formed under NT appear to resist damaging effects of rain and wind more effectively than aggregates formed under CT. In a study of soil samples of different-textured soils (loam, silt loam, sandy loam, and silty clay loam) collected from long-term plots (13-37 years) in Nebraska, Ohio, Michigan, and Kentucky, it was found that particulate organic matter associated with aggregates is affected by tillage, with a slower rate of macroaggregate turnover with NT (Six et al. 1999). The slower turnover rate helps preserve SOM as it is less subject to losses by erosion or decomposition, and the greater number of macroaggregates retained beneficially influence soil quality (better porosity, improved air and water exchange, and increased infiltration).

The benefits of NT appear to increase with time, evidently as the increase in SOM reduces erosion and improves soil structure. A 25-year study of NT conducted on four different Ohio sites indicated corn and soybean yields increased with time on all well-drained soils. Negative aspects of NT on poorly drained soils were largely eliminated by rotations, use of disease resistant varieties, and long-term maintenance of NT (Dick et al. 1991).

OTHER BENEFITS OF REDUCED TILLAGE

Reduced tillage has the potential of increasing crop production, not only because of an increase in SOM, but because much of the OM added remains on the surface as mulch. We have seen in Chapter 8 that some of the primary advantages of natural organic mulches over plastic mulches are (1) better soil moisture; (2) insulation of the upper soil, which reduces the possibility of excessively high or low soil temperatures; (3) improved aggregation and soil structure; and (4) improved porosity and aeration resulting from improved structure. These advantages also hold for RT over CT since most of the OM produced with RT remains on the surface as mulch. Another advantage of mulch, namely limiting erosion, although not appreciably better for organic over plastic mulch, is an important advantage of RT as compared to CT, which tends to bury most of the OM. Examination of the benefits of RT could help evaluate it as a means of increasing crop production and thereby aid in the sustainability of agriculture.

Soil Moisture

A major benefit of RT is the increase of soil moisture due to better infiltration and storage of rainfall or irrigation, and reduced evaporation. Reduced tillage, which allows better surface coverage with organic materials, provides greater infiltration. The surface roughness helps trap rainfall, allowing it to infiltrate the soil rather than being lost to runoff. Changes in the soil caused by the surface OM, such as reduced capping, greater amount of macroaggregates, and increased channels from earthworm activity, favor the ready movement of water into the soil. Although CT soils have greater intake soon after tillage than soils subject to reduced tillage, this advantage soon declines, evidently due to surface sealing. The increase in infiltration caused by CT will vary depending upon the amounts of OM and the percentage of soil coverage, as well as the type of soil.

Soil moisture increases under RT through reduced evaporation, as the surface OM protects the water from evaporation by reducing soil temperatures and protecting the water from wind. The improvement in water storage through RT can be substantial, at times providing twice as much water as CT. The differences in soil moisture usually are apparent through the entire growing season, and can be very helpful during hot, dry summers when moisture can become limiting.

As has been pointed out, the increased moisture with RT is not always beneficial. A cool, wet soil in early spring can delay planting and early development of several crops enough to adversely affect yields. Strategies designed to overcome these negative aspects of conservation tillage are covered in the later sections Stubble Mulch, Zone or Strip Tillage, and Ridge Tillage.

Lower Soil Temperatures

Surface soil temperatures tend to be appreciably lower with RT than with CT. The differences are greatly accentuated with full radiation during the heat of the summer months, at which time the reduction in soil temperature can have beneficial effects on soil biota and crops. However, the reduction in temperature can have harmful effects during spring and fall when soil temperatures may be too low with RT. Use of RT despite this deficit is also covered under Stubble Mulch, Zone or Strip Tillage, and Ridge Tillage.

Aggregation and Structure

Conservation tillage tends to increase the number and stability of soil aggregates. Soil aggregate stability helps provide long-lasting favorable structure. The aggregate formation and its stability increase in RT probably because the extra organic matter furnishes the energy for microorganisms that provide the mucus or cementing agents capable of binding the individual soil particles into aggregates. The improvement of aggregation and structure increases with additional years of RT.

Aggregation, by cementing together small soil particles into larger units, benefits soils in many ways. The benefits increase if the aggregates are relatively stable so that they resist the action of water, thereby lessening erosion and compaction, while providing better aeration and allowing better movement of both air and water for longer periods. The improvement in air and water movement make better aggregated soils easier to manage. Better soil aggregation and structure probably explain why soils with greatly reduced tillage are still able to provide sufficient aeration and drainage despite very little soil disturbance. Cover crops, one of the means for providing extra organic matter, play an important role in better aeration and drainage by increasing the number of water-stable aggregates. For example, the inclusion of cover crops in no-till sorghum production has been shown to increase water-stable aggregates (Table 9.1).

Porosity and Aeration

As SOM declines, much of a soil's porosity may be lost. Reduced tillage improves soil porosity by increasing SOM and decreasing implement traffic. The organic matter increases soil porosity by aggregating fine particles into porous granules, and compaction is reduced as a result of using lighter machinery and reducing the number of trips over the field. Porosity is further aided by earthworm activity, which is substantially increased by more favorable moisture conditions due to organic matter on the soil surface and by reduced disturbance from tillage machinery.

The increased porosity is usually reflected in better growth, fostered by the increased water infiltration and the better movement of

TABLE 9.1. The increase of water-stable aggregates after three years of no-till sorghum production as influenced by the addition of carbon from cover crops

Cover crop	Annual carbon input from cover crop (lb/acre)	Soil organic carbon* (%)	Water-stable aggregates (%)
None	–	0.85	28.9
Wheat	812	0.89	32.6
Hairy vetch	1,103	1.02	36.7

Source: Hargrove, W. J. 1990. Role of conservation tillage in sustainable agriculture. In J. P. Mueller and M. G. Wagger (Eds.), *Conservation Tillage for Agriculture in the 1990's.* North Carolina State University Special Bulletin 90-1. Raleigh, NC: Department of Crop Science and Soil Science, North Carolina State University.
*Multiply by 2 to obtain approximate percentage SOM

air and water in the soil. This tends to provide for an excellent balance between oxygen and water, preventing shortages or excesses of either components so vital for plants and soil biota.

Most plants and the soil biota cannot function without sufficient O_2 in the root zone, as it is necessary for respiration by which plant roots and most soil organisms obtain the energy to carry out vital functions. Carbon dioxide released in the soil by the respiration process must be exchanged for fresh air from the atmosphere to maintain the vital processes. The exchange of air high in CO_2 for air with sufficient O_2 is largely dependent on the porosity of the soil. There must be enough of a continuum of large pores, not only for the exchange of air so that sufficient oxygen is obtained, but also to drain excess water, which may block air from reaching the roots or soil organisms.

Conventional tillage introduces large amounts of air rich in O_2 into the soil by various implements used to work the soil. The moldboard plow is a superb implement for aerating the soil. It also does an excellent job of increasing infiltration, enabling the ready movement of water into the soil. Unfortunately, these advantages are not long lasting. The greatly increased aeration of the soil leads to the rapid decomposition of SOM, and in a relatively short time depending a great deal on climatic conditions, CT soils lose much of their porosity. The result usually is insufficient O_2 and at times insufficient or excess water.

Erosion

One of the great benefits of reduced tillage is its reduction of erosion. By reducing erosion, RT offers one of the major means of maintaining sufficient organic matter so that soil productivity can be maintained and agriculture sustained. The importance of reducing erosion could become even greater as the need to supply food for increased populations could force cultivation of soils that are more subject to erosion. It is a well-known fact that erosion can quickly reduce the productivity of shallow soils but has much less impact on production of deep-profile soils. Most of the deep-profile soils are already in production, leaving shallow soils, many of them on elevated slopes that are more subject to erosion, to supply extra food. Added to this list of endangered soils are the many acres formerly in rainforest now being made available for cultivation. The nature of rainforest soils and the large amounts of rainfall to which they are exposed make them subject to serious damage by erosion with quick loss of productivity unless extensive conservation tillage is used.

Conservation tillage tends to reduce the splash effect, essentially interrupting the chain of events before excessive damage is done. Conservation tillage also reduces erosion because the residues left on the surface tend to provide a more porous surface that is relatively free of crusts and is less apt to be sealed by water action. The lack of crusts tends to provide a soil that is open to infiltration at the beginning of the water addition. The reduction in sealing tends to provide for better continuing infiltration as irrigation or rainfall continues.

Wind Erosion

Conservation tillage reduces wind erosion by several means. More stable aggregates favored by reduced tillage are more resistant to detachment and movement. Organic matter provides a rough surface that lessens erosion by deflecting the force of the wind and shielding the soil particles from detachment. The surface OM also tends to keep the surface soil moist, and moist soil is less subject to wind movement. In fact, one method of reducing wind erosion is to moisten the soil prior to expected destructive winds, but irrigation is not available in many areas of extensive wind damage. Also, conservation tillage favors the use of vegetative covers (cover crops, living mulches) to reduce both wind and water erosion.

Soil Conservation and Erosion Losses

As mentioned in Chapter 1, the harmful effects of erosion were demonstrated during the extended drought of the 1930s, and efforts to lessen the damage were made by use of soil conservation practices promoted by the Soil Conservation Service, which was established in 1935. Practices promoted included windbreaks, terraces, and planting on the contour. Considerable headway was made in reducing erosion, but some reversals occurred as policies changed in the 1970s to favor increased production.

Fortunately, there again is an emphasis on erosion control. The Conservation Reserve Program, established as part of the 1985 Food Security Act, promoted land removal from crop production with conversion to long-term perennial vegetative cover. The USDA prepared 10- and 15-year contracts during the period 1986-1992, whereby about 35 million acres of highly erodible land, about 10 percent of total U.S. cropland, was removed from cultivation for a designated period and placed in perennial vegetative cover of grasses and forests. The system has been voluntary with accepted acreage receiving annual rentals and some assistance in establishing desirable covers.

The system has been successful, as there has been a major reduction (about 25 percent) in soil erosion of croplands in the United States. Soil eroded annually from cropland in the period 1982-1992 has fallen from about 2.75 to about 1.88 billion metric tons. Only about 60 percent of this reduction has been attributed to practices initiated under the Conservation Reserve Act and the remainder to conservation tillage, which has expanded during this period. Unfortunately, about one-third of the cultivated cropland in the United States is still losing over 4 tons of soil per acre per year to erosion. This amount is considered the maximum loss that can be sustained without serious production loss on most soils (Brady and Weil 1999).

The contracts to remove highly erodible soils from production have begun to run out and will be completed in about five years. It is problematical whether they will be renewed, making it all the more desirable to greatly increase conservation tillage along with adding extra OM, not only on highly erodible soils but on all cropland in order to reduce erosion to acceptable levels. Because conservation tillage can reduce the costs of producing a crop without sacrificing yields, it may be the only way to sustain sufficient agricultural production during this period of low farm prices to meet the new pressures for increased production looming in the not-too-distant future.

REDUCTION IN COSTS

Besides the benefits, conservation tillage has the great advantage of reducing production costs. Less power is needed to prepare soils for conservation tillage. Only about one-third to one-half of the horsepower of CT is needed for most reduced tillage, allowing smaller tractors to be used. Because of fewer operations, less time is needed for RT. It has been estimated that it takes 42 minutes of disking, ripping, field cultivating, and planting to prepare an acre of cotton using conventional tillage but only 6 minutes using NT. The fewer operations and smaller tractors consume less fuel, with estimated savings of 3.5 gallons of fuel, per acre (Bradley 2000).

NO-TILLAGE

The designation "no-tillage" may be a misnomer. Although it is now possible to grow crops with almost no tillage, some tillage is usually necessary to start the crop. The old crop can be destroyed by herbicides or by mechanically shredding or rolling so that it is not an impediment in establishing the new crop, but often it is necessary to loosen the soil enough so that seeds can be properly placed and firmed in order for the seed to germinate and the seedling to become established. Opening the soil with a coulter and ridge plow just ahead the seeder can in most cases provide a suitable environment for that start.

No-tillage is well suited for soils with good drainage that warm up quickly in the spring, for crops planted in midseason, or for fall-planted grains. It is also advantageous for some problem soils. No-tillage is well suited for sloping lands where it can markedly reduce erosion or for rocky fields where cutting coulter and planter double disk openers can override the rocks until normal soil conditions prevail.

Special planters equipped with a variety of coulters have been developed to cut through the surface debris and enable effective establishment of the new crop. Coulter selection is vital to provide proper cutting action, which allows planter openers to move unimpeded through the soil opening and place the seed at the proper depth. The narrow furrow opened needs to be closed and firmed in order to provide proper contact with the seed to promote even and rapid seed emergence. The residues remaining on the surface reduce the stand of emerging weeds, which can be still further controlled by use of herbicides.

The system has great advantages. Greatest responses to NT are on well-drained, low-SOM soils where a great improvement in SOM close to the soil surface has a substantial effect on upper soil structure, allowing better water and air management. Responses to NT appear to result more often on light-colored loam soils that tend to crust easily rather than on dark-colored silty-clay loams, which do not tend to crust but rather shrink while drying. Such shrinkage is beneficial to the latter soils, as it tends to form cracks that allow air and water penetration.

In well-drained soils, NT allows greater development of channels left by earthworms as the surface residues help provide better moisture conditions. The increase in channels and the improved water infiltration resulting from worm activity is cumulative since channels are not periodically destroyed by soil preparation machinery.

Poor Responses with No-Till

Although NT usually tends to provide better SOM retention, there are exceptions. The greatest increases in SOM with NT appear to be in soils with more than 2 percent slope, at least partly due to decreased erosion. A study in Illinois revealed that conservation tillage (which includes NT) did not increase SOM storage in the upper 12 inches of soil. This was attributed to little or no impact on soil erosion due to the low relief and poor drainage of the region. The small increase in SOM in the upper 2 inches of soil evidently was due to redistribution rather than an overall increase, since lower soil layers had less SOM in all but sandy soils (Needelman et al. 1999).

Restricting tillage almost completely, as is done in NT, is not suitable in all situations. No-till is not well suited for poorly drained soils. Infiltration, which is usually greatly improved with no-till on well-drained soils, is usually still poor on soils with clay pans or other impediments to water movement. No-till keeps these soils too wet, as the residues on the surface tend to restrict evaporation. Although NT may be associated with the greatest increase in SOM in most soils, the inadequate drainage usually associated with fine-textured soils limits the usefulness of NT. Under conditions that delay planting due to excess moisture and lower soil temperatures, crop production and crop residues will be greater with CT.

As was pointed out in Chapter 8, the mulch residue also insulates the soil from the warming spring sun. The combination of evaporation reduction and insulation help keep poorly drained soils, and even moderately drained soils, too wet, delaying spring planting. Delaying spring planting can reduce yields and often quality as well if time is limited between last killing frost in the spring and first killing frost in autumn. For example, planting corn in the Midwest in early May tends to provide optimum yields. Delaying the planting beyond about May 20 can seriously cut into yields.

In much of the northern corn belt, farmers trying to use no-till have the dilemma of delaying the planting until there is insufficient time to ripen the crop or planting it on time in undesirable soil that probably will provide a poor crop. A poor crop is almost assured, because germination and early growth will be slowed by the unfavorable soil conditions. The cold wet soils tend to increase disease because stress conditions usually accentuate disease problems.

The additional moisture associated with NT also appears to be a detriment for wheat and barley production, especially if grown on clay soils (Rasmussen et al. 1997; Legere et al. 1997).

Soil Compaction

Marked reduction in tillage would lead one to believe that soil compaction might become a serious problem in conservation tillage. Although problems do exist, the severity is not as great as expected in some circles, and it can be corrected. There may be several reasons for this; the obvious possibility is that increased organic matter reduces compaction from foot and machinery traffic. Since organic matter allows better drainage, soils in conservation tillage are less likely to be too wet in the fall, when soil compaction from harvest traffic produces some of the worst problems.

Despite the advantages of extra SOM in conservation tillage, compaction over time can limit performance of no-till for certain soils.

Much of the compaction results as soil particles settle due to gravity or from the use of various farm machines. Compaction resulting from farm equipment is usually worse in wheel traffic locations. Despite the benefits of soil residues, natural and machine-induced compaction can so compact soils as to warrant periodic plowing or other tillage. Although not as useful as a moldboard plow in overcoming soil compaction, use of a chisel plow ahead of the planter can correct many soil compaction problems. Used on a regular basis, it can prevent compaction problems on most soils from becoming serious enough to warrant use of the moldboard plow.

Stratification

No-tillage has other disadvantages. Limestone and slowly soluble fertilizers, such as superphosphates applied in NT, tend to remain close to the surface. In time, this can result in acid and infertile soil layers relatively close to the surface (4-8 inches), which may limit the development of a more effective root system. Some of the harmful effects appear to be offset, at least in some soils, by more effective roots close to the surface resulting from better moisture and temperature situations.

Probably because of the poor performance of NT on clay soils or other soils with poor drainage, especially in zones with limited growing seasons or zones seeking early markets, some growers have been hesitant to use NT. But rather than give up on the benefits of NT, many of them have worked out modifications whereby some cultivation tends to overcome temperature and moisture problems. Removing the residues from the center of the row (Kasper et al. 1990) or using zone tillage by in-row loosening with fluted covers (Pierce et al. 1992) improves grain production with no-till for some of these areas.

REDUCED TILLAGE

Forms of reduced tillage that allow some additional tillage on soils poorly adapted to NT will usually do better than CT, partly due to lower costs and improved SOM associated with NT. The additional tillage helps overcome some of the limitations imposed by poor drainage and the poor distribution of certain soil amendments, pest problems, and unacceptable compaction associated at times with NT. The extra tillage can consist of plowing the soil after a few years of NT or by using one of the methods of reduced tillage. Some of these variations of RT that provide varying degrees of SOM improvement, but still offer the possibility of sustainability despite their unsuitability for NT, are briefly discussed here.

Stubble Mulch

Stubble mulching allows for most residues of the previous crop to remain on the surface, yet rectifies many of the problems associated with no-tillage for poorly drained soils. Residues are spread evenly over the land and incorporated with implements that leave most of the residues on or near the surface (see Table 8.3). The use of V-shaped plows with blades 48-72 inches apart operating at 3- to 5-inch depths are especially helpful since they tend to retain practically all of the residues at the surface. The aim is to cover at least 30 percent of the surface. The residue on the surface tends to stabilize the soil, reducing erosion, improving infiltration, and permitting a suitable seedbed for relatively early planting unless the soil is very poorly drained. Crops can be planted the following spring using special planters that cut through the trash. Residue coverage tends to suppress early weed development.

Zone or Strip Tillage

Reduced tillage that limits tillage to strips offers several advantages. A major advantage of the system lies in having a zone that can be made ready for planting in the fall when moisture conditions may be more favorable, while the remainder can be devoted to mulch cover with all its attending benefits. It also allows for soil preparation ahead of planting, thereby spreading labor and movement of fertilizers. The early preparation provides a strip about 4-6 inches wide, 6-8 inches deep, and 1-2 inches in height that is drier and warmer than the undisturbed soil. Unlike some strip cropping described earlier in Chapter 6, these strips are relatively narrow, with only 4-6 inches cultivated and the remainder of the row (24-32 inches) left as undisturbed residue or cover crop. Planting can proceed rapidly, and the planted seed can get off to a quick start in the small prepared strip. Corn production following wheat or on heavy textured, poorly drained soils that are slow to warm up in the spring is favored by zone tillage in the fall. The practice also allows for fall application of P and K with little or no nutrient loss (Bruulsema and Stewart 2000). In the northern corn-belt, the early start can make the difference between success and failure.

Uncultivated strips between limited tillage rows confine wheel traffic, reducing the amount of compaction in the row. If planted on the contour, the uncultivated strips also allow orderly water runoff with limited erosion. Although more costly than NT, considerable savings in machinery, fuel, and time can be made compared to CT. Because of the savings and flexibility in planting preparation, there has been wide use of strip tillage, particularly on poorly drained soils.

Strip tillage on some cold, wet soils in Minnesota produced about the same corn yield as CT, both of which were about 10 bushels more per acre than were produced with NT, providing that soil tested high for phosphorus or a phosphate starter was used. There was no difference in yield with the different systems if soils tested low for P unless a phosphate starter was used. A one-pass system using cultivation or disking in the spring without primary tillage in the fall gave the same yields as CT or strip till (Randall et al. 2001).

Ridge Tillage

Zone or strip tillage on near-level soil may still cause problems on soils that do not drain well, especially if prepared soils are subject to rain. In such case, the prepared strip may become too wet for early planting or for emergence and growth. Such problems can be avoided by preparation of a bed or ridge on which the primary crop is planted. Ridge tillage also limits cultivation to relatively narrow strips but, unlike strip tillage, planting is done on a ridge about 4-6 inches high.

Ridge planting has gained ready acceptance, especially in areas where soils may stay wet and cool late in the spring and commonly do not do too well with NT. It has the advantage of lower costs than moldboard plowing, but on well-drained soils may provide a lower yield of corn and soybeans than is obtained with a well-managed moldboard plow operation. Its best use is on medium- to heavy-textured soil that is level or slightly sloping. In no case should it be used up and down the slope if the slope is greater than 3 percent, unless the ridges are contoured.

The alleys between ridges are used for farm machinery traffic, greatly eliminating compaction in the seed zone and almost all of the compaction in the root zone. If the rows are on the contour, ridge tillage offers erosion control on relatively steep soils. Contoured ridge tillage has provided successful corn production on acid steepland-oxisols in the Philippines that are highly subject to erosion. The system, either alone or in combination with contour grass barrier strips, produced greater corn yields than contour moldboard plowing at much lower costs (Thapa et al. 2000).

Ridge tillage, with its ease of planting the same rows each year, lends itself to wide (10-20 ft) strips, in which corn and soybeans are planted in alternate strips. Soybeans may get some shade in the first few rows adjacent to the taller corn, which may be partially reduced by planting a strip of small grains between the strips of corn and soybeans.

The multiple strip cropping offers the advantage of larger amounts of residues and therefore less erosion, especially if furrows between ridges (alleys) in the small-grain strips are also planted to small grains. There is some question whether these alleys should be planted, because in wet years yields may be reduced rather markedly. The current sense is that the practice is worthwhile, partly because in dry years the increased yields of alleys tend to offset the lower production on the ridges.

Although ridge tilling eliminates much of the tillage associated with conventional tillage, some tillage is necessary. The first operation usually consists of planting corn or soybeans in the previous year's ridges. It usually is necessary to sweep aside some trash ahead of the planter and possibly add fertilizer and herbicides at that time. The second and often final operation occurs when corn and soybeans are about 20 inches tall. Sufficient soil from the alleys is thrown up near the base to rebuild the ridges for the following year's crop.

Some adjustments in machinery are necessary to utilize ridge tillage. If heavy residues are present, a stalk reduction program can aid in soil preparation. Either a rotary tiller or sweep tiller with trash rods is used in flat soil preparation. Fertilizers are applied in late fall or early spring ahead of the planting. The rotary tiller is centered ahead of each row planter and can incorporate herbicides and fertilizers. The sweep tiller with trash rods or a disk row cleaner opens a small area, allowing the planter that follows to move relatively unimpeded.

A cultivation using disks to move soil away from the plant and then back again plus a sweep in the row middle is usually enough to control weeds.

Necessary equipment for ridge tilling consists of a cultivator capable of making ridges and a planter designed to plant on ridges. The cultivator can use sweeps or heavy-duty disks to build a ridge 4-6 inches high. A stalk chopper is also highly desirable for growing corn on ridges. Existing corn stalks need to be chopped in before planting in corn residue. If stalks are chopped in the fall, the residues help protect the soil from erosion while good soil structure tends to form beneath them. Fertilizers are broadcast before ridges are formed or anhydrous ammonia is knifed in between ridges. A sweep with trash bars is used at planting to remove 2-4 inches from the surface, allowing rapid planting. Cultivation using disks is used to move soil away from the plant and then back again. Sweeps added to heavy-duty disk cultivators rebuild the ridges at layby.

Cover crops of ryegrass and vetch have been used in the Northeast. Seeded by air in late summer or early fall, seeds tend to bounce or slide off ridges to germinate in the row. The cover crop is killed by herbicides in the spring. Sweeps with trashbars can be used to clean the surface and rebuild the bed for replanting.

Zone and ridge tillage can be expected to give excellent erosion control. Stubble mulching also can reduce erosion, but the extent is closely related to amounts of residue and how much of it remains on the surface. Although providing less cover on the surface than no-till, these modified programs provide much more cover than various forms of CT (moldboard plow, chisel, or disk).

PEST PROBLEMS WITH CONSERVATION TILLAGE

Pest problems, noted primarily with NT, can be aggravated in all forms of CT. Certain pests can benefit from accumulation of residues at the surface and lack of soil turnover. Residues can provide overwintering sites for insects, disease organisms, nematodes, slugs, and small invertebrates. Additional moisture, usually present with CT, may also increase weeds.

Insects

For example, damage to corn may be greater when grown with reduced tillage than with CT because of damage from armyworms (*Pseudaletia unipuncta* Haworth), black cutworms (*Agrotis ipsilon* Hufnagel), stalk borers (*Papaipema nebris* Guenee), Southern corn billbugs (*Sphenophorus callosus* Oliver), corn root aphids (*Anuraphis maidiradicis* Forbes), cornfield ants (*Lassius allenus* Forster), western corn rootworms (*Diabrotica virgifera* LeConte), northern corn rootworms (*D. longicoruis* Say), southern corn rootworms (*D. undecimpunctata howardi* Barber), wireworms (*Melanotus* spp., *Conoderus* spp. and *Ludius* spp.), sugarcane beetles (*Euetheola rugiceps* LeConte), seedcorn maggots (*Hylemya platura* Meigen), and the grasshoppers—which also affect other crops—redlegged (*Melanoplus femurrubrum* DeGeer), differential (*M. differentialis* Thomas), and migratory (*M. sanguiipes* Fabricius). Seeds appear to be particularly vulnerable to the increased damage associated with conservation tillage. Besides wireworms and seedcorn maggots, the seedcorn beetle complex (*Stenolophys lecontei* Chandoir and *Clivinia impressifrus* LeConte) can cause serious damage at times (Sprague and Triplett 1986).

A combination of increased insect carryover and several viruses have increased maize chlorotic dwarf and maize dwarf diseases of corn grown under conservation tillage systems.

Not all of the insect infestations increase with conservation tillage. A number, such as European corn borer (*Ostrinia nubilales* [Hubner]), sorghum midge (*Contarinia sorghicola* [Coquillett]), and fall armyworms (*Spodoptera frugiperda* J. E. Smith), appear to be equally damaging with both systems. Some insect pest damage, such as that caused by the Mexican bean beetle (*Epilachna varivestis* Mulsant), lesser cornstalk borer (*Elasmopalpus lignosellus* Zeller), and mint root borer (*Fumibotys fumalis* Guenee), may be reduced with conservation tillage (Sprague and Triplett 1986).

Crops other than corn may be damaged more severely by insects when grown with conservation tillage. Some typical crops suffering additional insect damage are cotton damaged by tarnished plant bugs (*Lygus lineolaris* Palisot de Beauois) and bollworm/budworm (*Heliothus* spp.), wheat affected by Hessian fly (*Mayetiola desstructor* Say), sorghum attacked by the sorghum midge (*Contarinia sorghi-*

cola Coquillett), and sorghum, soybeans, and small grains damaged by corn earworm (*Heliothis zea* Boddlie) (All and Musick 1986).

Disease

A number of plant disease organisms are a concern because they can overwinter in residue promoted by conservation tillage. They include southern corn leaf blight (*Helminthosporium maydis* Nisakado), corn anthracnose (*Colletotrichum graminicola* Ces.), yellow leaf blight of corn (*Phyllosticta maydis* Arny and Nelson), corn eyespot (*Kabatiella zeae* Narita and Hiratsuka), tanspot of wheat (*Pyrenophora trichostoma* Fr Fckl), cephalosporium stripe of wheat (*Cephalosporium gramineum* Nsikado and Ikata), root rot of wheat (*Cochiobolus sativus* Ito and Kurib), Holcus leaf spot of corn (*Pseudomonas syringae* Van Hall), Goss's bacterial wilt of corn (*Corynebacterium michiganense* ssp. *nebraskense*), septoria glume blotch and septoria blotch of wheat and barley (complex of *Septoria nodorum, S. tritic,* and *S. avennae* f. sp. *triticea*), southern leaf blight of tomato *(Sclerotium rolfsii),* brown spot of corn *(Physodermam maydis),* verticillium wilt of cotton *(Verticillum dahliae),* bacterial blight of soybeans (*Pseudomonas syringae* pv. *glycinea*), halo blight of beans (*Pseudomonas syringae* pv. *phaseolicola*), bacterial blight of cotton (*Xanthamonas campestris* pv. *malvacearum*), tomato canker (*Clavibacter michiganensis* ssp. *michiganensis*), and cereal black chaff (*Xanthamonas campestris* pv. *translucens*) (Boosalis et al. 1986).

Weeds

The almost complete lack of tillage with NT allows some weedy perennials to become serious problems. Plants such as poison ivy, horse nettle, trumpet creeper, and even tree seedlings can become established. In addition to these plants that may be difficult to control, a number of others have been listed as problem weeds for conservation tillage. The kind of weeds vary with different crops, since the weed's competitiveness varies with the primary crop, and different herbicides are available for different crops. Weeds such as johnsongrass, bermudagrass, hemp dogbane, wild proso millet, quackgrass, itchgrass, switchgrass, shattercane, purple nutsedge, and jointed goatgrass were considered problem weeds in at least two of the five crops (corn, cotton, sorghum, soybeans, and wheat), grown under no-tillage or surface tillage (Triplett and Worsham 1986).

In CT, weeds are controlled by cultivation or herbicides. Controlling weeds by cultivation is reduced to varying degrees with different types of conservation tillage. Herbicides can still be used effectively, but other approaches are also available (see Chapter 10).

Nematodes

As indicated earlier, nematode problems may be difficult to control with conservation tillage if the pest is well established, but conservation tillage tends to keep major problems from developing. The highly increased biological diversity probably is an important factor in keeping nematode problems in check. The greater use of cover crops also may contribute to control, as several cover crops, such as sorghum-sudangrass hybrids, marigold, showy crotalaria, sunn hemp, velvet bean, and cereal rye are capable of limiting growth of at least one plant-damaging nematode. Cereal rye is effective against root knot, reniform, and stubby root nematode (Phatak 1998).

Conservation tillage usually employs rotations, which may contribute to reductions in nematode damage. Rotations with crops not susceptible to root knot have been used for many years to reduce damage to susceptible crops. The nonsusceptible crops have varied with different primary susceptible crops. Alfalfa, small grains, and sorghum have been used to combat root knot in cotton, caused by *Heterodera marioni* (Cornu) Goodey.

Peanuts, velvet beans, crotalaria, cereals, and forage grasses have been rotated with many different vegetable plants susceptible to root knot, caused by the same organism. Alfalfa, sweet clover, beans, peas, potatoes, cereals, and many vegetable plants, except beets and all crucifers (cabbage, cauliflower, etc.) have been rotated with sugar beets to avoid serious damage from the sugarbeet nematode (*Heterodera schatii* Schmidt).

The length of time for which the nonsusceptible crops need to be grown before the susceptible plant can be brought back into the rotation varies with the severity of previous damage, but usually is two to three years. With sugar beets, a period of five to six years of nonsusceptible crops is necessary before sugar beets can be grown for a year.

Rotations have been shown to be effective in reducing populations of soybean cyst nematodes (*Heterodera glycines* Ichinohe), with benefits being greatly increased with no-till over CT (Table 9.2). Although not noted on the table, nematode control associated with no-till was limited to seasons of limited moisture due to lack of rain or irrigation.

Rodents and Birds

Mice can burrow in the slot opened by planters in reduced tillage systems, eating the seeds for some distance in the row. In orchards, conservation tillage favors mice, which often burrow near young trees and effectively destroy enough bark and roots to seriously interfere with tree vigor so that it will develop poorly or winter-kill.

Bird problems at harvest appear to be about the same for CT and conservation crops, but poor harvesting practices that result in seed losses may increase damage in newly planted conservation tillage fields.

TABLE 9.2. Effects of rotations and tillage on populations of soybean cyst nematodes

	Tillage (per pint of soil)					
	Conventional			No-till		
Crop sequence	Cyst	Eggs	Juv.*	Cyst	Eggs	Juv.*
Continuous soybeans	150	144475	305	52	3675	80
Soybeans after corn	122	11388	163	57	4513	105
Soybeans after corn and wheat	126	12300	275	15	1625	8
1 year corn after soybeans	1	13	1	4	300	0
1 year corn after wheat and soybeans	0	0	3	2	75	8
2 years corn after soybeans	0	0	0	2	75	0

Source: Koening, S. R., D. P. Schmitt, and B. S. Sipes. 1990. Integrating conservation-tillage and crop rotation for management of soybean cyst nematode. In J. P. Mueller and M. C. Wagger (Eds.), *Conservation Tillage for Agriculture in the 1990's: Proceedings of the 1990 Southern Region Conference, July 16-17, 1990.* Raleigh, NC: Department of Soil Science, North Carolina State University.
*Juvenile stage

Allelopathic Effects

There have been reports of reduced yields of crops in conservation tillage because the growth of the primary crop was adversely affected by allelopathic chemicals in the previous crop's residues. Reports of reduced corn yields following wheat have been fairly common, but an experiment to evaluate these reports failed to show any detrimental effects from wheat straw. In fact, when wheat straw was removed from certain experimental plots, corn height and yield was less than if the wheat residue was left undisturbed or residues were doubled. If there were any restrictions in yields due to allelopathic chemicals in the wheat, they were more than overcome by the beneficial effects of wheat residue mulch (Varsa et al. 1995).

CORRECTIVE MEASURES

The actual problems with conservation tillage, especially those related to pests, appear to be much less serious than anticipated. In many cases, the overall need for insecticides and fungicides often is appreciably less with reduced tillage than CT. The reduction in pesticide use has been attributed to the greater biological diversity under conservation tillage. The presence of beneficial organisms that keep insects and disease in check are known to increase with conservation tillage, but this increase appears to vary with different locations, soils, climate, and the extent of protective cover for the beneficials. Where conservation tillage does not provide suitable protection, various corrective measures can be used.

No-tillage can be used on some poorly drained soils because some of the negative impacts of NT may be overcome with better disease-resistant varieties and crop rotations. In time, continued use of NT also seems to eliminate the negative response to NT (Dick et al. 1991). Although these approaches may be used with NT, most growers faced with poorly drained soils and/or short growing seasons have favored using a variation of NT (stubble mulching, strip or zone cropping, or ridge tillage) that provides some tillage to eliminate the problem of poor drainage or reduced growing season. Tillage may also be needed to better position limestone and phosphate, although proper placement at the beginning can go a long way in maintaining satisfactory yields. Compaction can be largely overcome by use of subsoilers.

Subsoilers can also be used to place phosphate and limestone deeply, two ingredients that move poorly into soils.

Although conservation tillage has the potential of increasing various pest problems, as has been pointed out, pest control may be better with some forms of conservation tillage. Despite this anomaly, there is little doubt that pest control problems and corrective measures for them are often far different with reduced tillage than with CT. Some of the primary differences and controls are examined more thoroughly in Chapter 10. The problems of pH and nutrient stratification as well as many other nutrient changes induced by conservation tillage are also examined more fully in Chapter 10.

The manner in which we can maximize crop production by combining increased additions of OM with reduced tillage to provide better pest control and nutrient availability, suitable soil aeration, less compaction, and reduced soil erosion is covered in greater detail in Chapter 11, "Putting It All Together."

Chapter 10

Changes Brought About by Conservation Tillage

CULTURAL METHODS

Conservation tillage produces considerable changes in a soil, some of which may require extensive modification of cultural methods. Most of the soil changes, due to the presence of large amounts of OM on the surface and overall increases of SOM, have been touched upon in the preceding chapters. Many of the changes in cultural practices are due to the accompanying reduction in tillage that they allow. Some of the cultural changes induced by increases in OM and SOM and the reduction of tillage that follows are summarized here.

Equipment for Conservation Tillage

Considerable changes are often necessary for equipment used for conservation tillage. Those used for strip and ridge culture are briefly covered in Chapter 9. Some of these changes, such as smaller or fewer tractors, already touched upon, result in considerable savings. Others require extra expense as old equipment is modified or needs to be replaced.

Organic crops (cover crops, sods, forages) used for conservation tillage will have to be terminated prior to planting the primary crop. Ideally, annual crops should be terminated when they have flowered but not set seed. The crop can be terminated by chemical or mechanical means. In Virginia, the ideal time for terminating cover crops prior to planting pumpkins is three to five weeks before planting, as this will allow leaching of allelochemicals from the terminated crop and restoration of soil moisture (Morse et al. 2001).

Chemicals methods using the contact herbicides glyphosate (Roundup Ultra) or paraquat (Gramoxone Extra) can be used to terminate a number of plants (cover crops, weeds, sods) but it may be necessary to apply the glyphosate three to five weeks and the paraquat 10-14 days ahead of planting the primary crop to get adequate control. It also may be necessary to make two or three applications for complete kill (Morse et al. 2001).

The need to apply glyphosate three to five weeks ahead of planting may make this approach incompatible with the need to produce a large biomass of cover crop for early crops. In such cases, it may be far better to allow the cover crop to grow another week or two and use paraquat or a mechanical method such as undercutting with rolling to terminate the cover crop. To avoid allelochemical effects of the cover crop residues, it may be desirable to sweep the residues away from the planting zone.

Mechanical procedures include tillage, mowing, and rolling. It may be difficult to terminate the crop with tillage unless it is extensive, which would nullify many of the benefits of conservation tillage. Sickle bar mowers tend to leave the residues in strips, and rotary mowers tend to pile the chopped residues in windrows. Neither method is as suitable as flail mowing or rolling for effectively terminating many cover crops, but flail mowing or rolling are not effective in terminating berseem, red or white clovers, or oats (Morse et al. 2001).

Flail mowers use double-edged knives mounted on a parallel rotor that finely cut the organic matter in small pieces that are evenly distributed on the soil surface, making it practical to plant through the residues with suitable coulters and trash sweepers.

Rolling has worked well in terminating cereal grain crops and some legumes, although rolling of cover crops is usually less complete than mowing. Rolling should be done in the direction of planting. Some of the equipment used for rolling is as follows:

1. Flail mower: Mature crop residues are effectively flattened by the roller and gauge wheel of disengaged mowers.
2. Grain drills can be equipped with coulters and cast-iron press wheels that are spaced 5 inches apart to roll cover crops.
3. Rollers filled with water, normally used for turf or construction, can effectively roll crop residues.
4. Roller-crimper drums fitted with blunt steel blades or metal strips welded horizontally and filled with water have been used to flatten cover crops, killing the crop by compressing and folding while leaving the stem intact.

5. Undercutter rollers with a V-plow sweep followed by a rolling harrow undercut cover crop roots while the residues are rolled flat on the ground.
6. Rolling stalk choppers (Figure 10.1) can be adjusted or modified to effectively chop stalks and evenly distribute high-residue cover crops (Morse et al. 2001).

Considerable changes in planters may be necessary to effectively plant crops in conservation tillage. The presence of crop residues or remains of cover crops and other OM can offer a challenge to proper planting, although it is considerably less for zone tillage, because moisture levels and compactness of the undisturbed soil are usually greater than with CT. The combination of trash and compactness makes it more difficult to deposit the seed at the correct depth and leave it in a zone with just sufficient moisture to germinate. Conventional planters often can be used for zone tillage with little or no modification but need considerable changes for no-till. No-till planters need to be fitted with coulters (Figure 10.2) to cut through the trash and double disk openers to place the seed at the proper depth and cov-

FIGURE 10.1. Buffalo Rolling Stalk Chopper suitable for rolling tall cover crops and leaving them as a uniform layer of residue on surface. (Photo courtesy of Dr. R. D. Morse, Virginia Polytechnic and State University, Blacksburg, Virginia.)

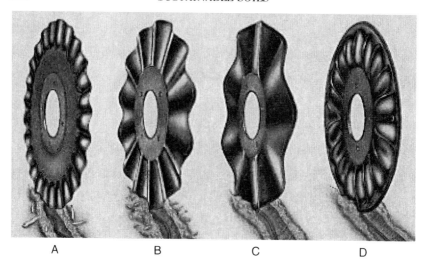

FIGURE 10.2. Several types of available coulters can be placed ahead of seeders to cut through residue and open small furrows suitable for seed placement. Coulter A is adapted for wet, spongy soil. Coulter B provides fine tilth even at low speeds. Coulter C opens a slot about 1¼ inch with low soil disturbance at very low speeds. Coulter D cuts through residue and penetrates well in wet and sandy soils, opening a slot about 1 inch wide. (Photo courtesy of Adrian Puig, Everglades Farm Equipment, Belle Glade, Florida, and Matthew A. Weinheimer, John Deere Des Moines Works.)

erage. A chisel plow positioned ahead of the coulter aids in loosening soil and preparing a friable zone for planting. Although conventional planters can be modified for no-till seeding, it may be worthwhile to use planters designed for no-till. Specially designed planters for no-till are becoming more readily available (Figures 10.3 and 10.4). Modifications to conventional planters can be made at costs of $100 to $500 per row unit, whereas new planters cost about $5,000.

Whether modifications will be made to an old planter or a new planter is purchased, the following items need careful attention:

1. A strong tool bar capable of supporting the necessary attachments
2. Pressure springs that can pull the planter toward the soil surface with enough force to keep the planter from coming out of the soil
3. Double disk opener units that can roll to avoid dragging trash while still cutting through residues and soil surface

4. Coulters or offset double disk openers that can cut through the trash and work the correct amount of soil for proper seed contact with the soil
5. Heavy-duty press or covering wheels (Figure 10.5) that are spring-loaded with enough pressure to cover seed for good soil contact and sufficient soil to protect seed from applied herbicides

Even with all the necessary accoutrements, planters need to be adjusted in the field to meet specific conditions. It usually is necessary to adjust springs and pressure settings for specific field conditions. Extra weights for the planters may be needed for the hard or dry soil conditions. Also, trash removers may be needed for special conditions when residues may be difficult to handle (Prepare planters for con-till, 2000).

FIGURE 10.3. Seeder capable of planting seeds and applying liquid fertilizer in 8 or 12 rows of residue mulch in varying terrain. (Photo courtesy of Adrian Puig, Everglades Farm Equipment, Belle Glade, Florida, and Matthew A. Weinheimer, John Deere Des Moines Works.)

FIGURE 10.4. (A) Planting pumpkin seeds in terminated rye cover crop. (B) Uniform, healthy pumpkin crop several weeks old growing in terminated rye cover crop. Note general vigor and lack of weeds. (Photo courtesy of Dr. R. D. Morse, Virginia Polytechnic Institute and State University, Blacksburg, Virginia.)

For some crops, it is desirable to send ground-driven stalk choppers ahead of the planters to facilitate easy movement. Usually these stalk choppers are more economical than the power take off (PTO) tools that shred the residues.

A chisel plow capable of opening the soil to a depth of 6-8 inches may also be needed on many soils in order to get enough air and water to move readily into the soil to provide for rapid seedling development. Multiple chisel plows can be mounted on a bar to open wide swaths of soil prior to planting (Figure 10.6), or a single chisel plow can be placed conveniently behind the coulters or double disk openers and before each seeder. Using a winged chisel plow for opening compact soils will provide a more suitable seedbed.

It may be more desirable to use vegetable and tobacco transplants instead of seed to start a crop using conservation tillage. Cucumbers,

A   B

FIGURE 10.5. Closing wheels after seeders can be varied to meet soil and mulch conditions. Standard double rubber-tired wheels (A) are designed for common fields. Cast iron wheels (B) may be needed for tough soil and heavy residues. (Photo courtesy of Adrian Puig, Everglades Farm Equipment, Belle Glade, Florida, and Matthew A. Weinheimer, John Deere Des Moines Works.)

cantaloupes, celery, peppers, tomatoes, watermelons, and tobacco can be readily transplanted and, because of early markets or savings in seed cost and early pesticides, can be more profitable than seeded plantings.

A new transplanter (Figure 10.7) is capable of planting various transplants in a variety of soils used for no-till, including those that are susceptible to erosion and drought, but not for those that are too wet. It has a special subsurface tiller in front of a double disk shoe of a transplanter. A narrow strip (2-5 inches) of soil is loosened by a 20-inch smooth coulter that is spring-loaded, mounted ahead of a fertilizer knife with a winged point. In one pass it can slice through heavy mulch, loosen a narrow strip in the row, band fertilizer on both sides of the row, install drip tubing 1-3 inches deep under the mulch next to plants, and place transplants (Aylsworth 1996; Morse et al. 1993).

An additional tool that is often needed for conservation tillage is a sprayer that can safely apply herbicides to terminate cover crops or weeds prior to planting and postplanting herbicides to control weeds after planting the primary crop. Sprayers should be hooded so that herbicides can be applied to weeds with minimal damage to the primary crop or to cover crops that need to be maintained to support beneficials. Cost of a postplanting new sprayer that is hooded or shielded varies from $4,000 to $10,000 depending on width of the unit and number of rows that can be covered.

FIGURE 10.6. Multiple chisel plow as a unit opens soil, leaving considerable residue on surface. The unit is designed to stay level, keeping sweeps or shovels at uniform depth. (Photo courtesy of Adrian Puig, Everglades Farm Equipment, Belle Glade, Florida, and Matthew A. Weinheimer, John Deere Des Moines Works.)

FERTILITY CHANGES

Modifications in a soil's fertility due to conservation tillage may make it necessary to alter fertilizer programs designed for CT. Again, most of the changes are due to the surface OM and increased SOM, but changes in pH are largely caused by the surface placement of N fertilizers and the inability to work liming materials into the soil.

FIGURE 10.7. Subsurface Tiller Transplanter developed by Dr. R. D. Morse and associates at Virginia Polytechnic Institute and State University, Blacksburg, Virginia. The transplanter is capable of loosening compact soil in a small band (2.5-4 in wide × 6-10 in deep), setting transplants in heavy residues, firming loosened soil around plants, applying needed fertilizers and pesticides, and laying drip irrigation tubing. (Photo courtesy of Dr. R. D. Morse.)

Changes in pH

Surface pH tends to be lowered with conservation tillage. Losses of bases, particularly Ca and Mg, which are accentuated under the more moist conditions, account for some of this loss in pH. The remainder is probably due to hydrogen ions formed as applied N derived from ammonia, ammonium nitrate, and urea fertilizers are oxidized by microbial action to form nitrates. The upper soil acidification is accentuated by placement of N fertilizers on or near the surface. The extent of pH drop is closely associated with the amount of N applied (Table 10.1).

Lower pH values may be less of a problem with conservation tillage than they are with CT. First, pH values usually drop less in conservation tillage than in CT due to the buffering effect of the larger reservoir of SOM. Second, although losses of Ca and Mg can cause problems with many plants, the presence of larger amounts of SOM

TABLE 10.1. Changes in soil pH values after ten years of corn grown with different N rates under no-tillage and conventional tillage

N Rate (lb/acre)	NT		CT	
	0-2 in	2-5 in	0-2 in	2-5 in
0	5.75	6.05	6.45	6.45
75	5.20	5.90	6.40	6.35
150	4.82	5.63	5.85	5.83
300	4.45	4.83	5.58	5.43

Source: Reprinted from *Soil and Tillage Research,* Volume 3, Blevins, R. L., G. W. Thomas, M. S. Smith, and W. W. Frye, Changes in pH values after 10 years of corn grown with different N rates under no tillage and conventional tillage, pp. 135-146, 1983, with permission from Elsevier Science.
NT = no-tillage; CT = conventional tillage

tends to mitigate some of the other ill effects associated with the lower pH. The reduction in available P with increased Al and the damage from excess Al to many sensitive plants is lessened by the larger amounts of SOM.

If there are sufficient losses of Ca and Mg and a reduced pH, it will be necessary to apply liming materials. The most effective method is to broadcast materials and then work them into the surface by plow or disk. This may not be a viable option because of decreased tillage. Fortunately, there is some evidence that the application of very fine liming materials to the soil surface can be sufficiently effective, eliminating the need for extra tillage. The effectiveness of surface-applied liming materials evidently is influenced by the moist soil surface favored by conservation tillage. The moist surface helps to keep Ca and Mg in solution and promotes a more efficient root system close to the surface.

Despite the advantages provided by the surface moisture in mobilizing the liming materials, the presence of considerable pH stratification may require some means of mixing liming materials with the topsoil. It may be possible to provide sufficient mixing by using RT rather than CT. Conventional tillage may undo much of the benefits of an accumulated layer of organic materials at the surface. Some forms of stubble mulching or the use of zone or ridge tillage may allow sufficient mixing of liming materials and topsoil to eliminate enough of the stratification to provide better root development. In zone or strip tillage, mixing materials into the soil is limited to a narrow strip 4-6 inches wide, but even this form of limited tillage can be helpful in mixing ingredients. Ridge tillage allows for more options in placing both limestone and fertilizer than zone tillage, and placement options are further increased with certain forms of stubble mulching.

Nutrients

Nutrients, especially those in slightly soluble forms, also may be concentrated near the surface with NT. The availability of these nutrients is usually increased by the good soil moisture and increased root activity near the surface. When excess nutrients may build up, corrections can be made by one of the forms of RT for correcting stratification of pH.

Availability and uptake of nutrient elements can be different in reduced tillage than in CT. Nitrogen is the most affected element, but there can be differences in other elements as well. Ample moisture under the surface OM favors less fixation of P and K as well as greater diffusion of several elements, but may contribute to losses of N by denitrification and leaching. On sandy soils, leaching losses of K, Ca, Mg may also be accentuated.

Nitrogen

There are appreciable differences in N availability with CT as compared to RT. Reduced tillage soils tend to have larger amounts of microbial mass and organic N, recycling N more effectively than CT soils. Because CT mixes and incorporates crop residues and other OM additions into the soil, microbial decomposition is hastened, and there is a quicker release of N from these sources than with RT, providing that soil moisture, O_2, and temperature are satisfactory. The combination of large microbial mass, large amounts of organic N, and rapid decomposition accentuated by the intimate mixing of OM and soil tends to provide ample N for many crops, especially if legumes have been incorporated.

The wetter conditions usually present with conservation tillage place it at a disadvantage during wet periods such as early spring in most of the United States, since decomposition necessary for the release of N will be slowed. The amount of N available from both OM and applied forms, mostly inorganic, also is less in the spring under conservation tillage because the wetter conditions favor the leaching of available N. But the wetter conditions can be an asset during the summer months when conditions favor uptake.

More N is also lost from conservation tillage during the very wet periods due to denitrification and at most other times if urea or urea-ammonium nitrate solutions (UAN) are surface applied. The greater loss of surface-applied urea is probably due to the presence of more enzymes with the larger amount of OM present under conservation tillage. The losses of N from urea, anhydrous ammonia, ammonium nitrate, or UAN solutions applied for conservation tillage can be reduced by soil injection (Mengel et al. 1982).

Dribbling ammonia-containing solution fertilizers can reduce the loss compared to broadcast, but if weather conditions following application are dry, there still can be considerable loss. The UAN or ammonium nitrate solution can readily be point injected by spoke-wheel injectors or through tubes behind chisel plows. For ridge tillage, a single application as a preplant point injection 1-3 inches from the crop row on the ridge tends to give excellent results.

Studies indicate that there are variations in the amounts of N recovered by plants grown with RT as compared to those grown with CT. Generally, CT will provide better uptake of N than no-tillage, especially during the first few years of no-tillage and if the amount of applied N is rather small. The differences in absorbed N between the two systems of tillage narrow with time, and the no-till actually can be ahead after several years if enough N is applied (Meisinger et al. 1985).

The exact cause for these differences in absorption with time and different levels of N is not known, but it is suspected that slower nitrification rates with NT in the early years, combined with higher rates of denitrification and leaching associated with additional moisture, may be responsible. With low application rates, a relatively large part of N may be tied up by microbial action because of wide C:N ratios. With larger applications, there is ample N to take care of the microbes and still leave enough N that is better absorbed under RT than with CT during the summer months because of the extra moisture.

The slowed release of N from organic sources under conservation tillage may require extra application of available N forms for rapid plant starts in the spring. Use of starter fertilizer N can be helpful at such times, but amounts need to be carefully monitored, as close placement is required for effectiveness and small excesses can cause damage.

Phosphorus

Availability of P is also altered by conservation tillage. The additional moisture present appears to increase the amounts available by increasing diffusion rates. Also, greater amounts of P are kept soluble by the extra OM associated with conservation tillage since SOM limits the amount of phosphate normally fixed by Al, Fe, and Ca. For example, humic acid formed from organic matter decomposition has been shown to inhibit formation of insoluble calcium phosphates, thereby increasing bioavailability of P (Grossl and Inskeep 1991). In calcareous soils, the additional acidity associated with surface OM, increases available P by solubilizing fixed phosphates. In acid soils, where the fixation of P is apt to be due primarily to high levels of Al, the extra organic matter reduces fixation by suppressing Al solubility.

The smaller amount of P fixed in the presence of high levels of OM allows greater discretion in P placement. Under conservation tillage, there is less need to band the P to keep it available, and it can be placed on the surface soil under the mulch. It has been suggested that placing P close to the seed with conservation tillage may be more important because the diffusion of P is very much affected by temperature. But a study with several tillage systems showed no advantage for applying starter phosphate for strip tillage compared to CT (Randall et al. 2001).

The reduced need for starter P (and possibly K as well) in no-till operations appears to be due to the large number of roots developed in the upper part of the soil, stimulated in part by the additional moisture and nutrients present. The stratification of several nutrients in the upper couple of inches in the soil occurs as the organic matter, largely confined to the surface, decomposes, releasing the nutrients close to the surface where cations and P remain due to a lack of mixing. The cations are also held in place by the exchange complex. This combination of upper stratification of nutrients and roots allows broadcasting of fertilizers if the soil test shows "good" levels of P and K at the surface.

It must be remembered that it may take several years of conservation tillage to produce conditions favorable for broadcasting P (and K) because little or none of these elements are fixed in soils with appreciable amounts of clay. Until that time, it probably would help if a good proportion of these elements was banded 4-6 inches deep and some of it added as starter fertilizer. Adding enough P and K as deep bands prior to starting conservation tillage should help prepare the soil to broadcast these elements.

The advantages in P availability due to activities of mycorrhizae, already covered in Chapter 5, are more readily utilized with conservation tillage than with CT, but their utilization may depend on modification of fertilizer practices. Application of fertilizers, particularly phosphates, and certain herbicides and insecticides appear to restrict the usefulness of mycorrhizae.

Potassium

Potassium diffusion is also benefited by the presence of ample moisture, which tends to favor conservation tillage over CT, but this advantage may be lessened by the poorer uptake of K in cool soils. The poorer absorption from cool soils may be responsible for the generally poorer uptake from soils in conservation tillage.

The availability of K is influenced by percentage saturation of the element in the exchange complex. From the senior author's experience, it appears necessary to have a minimum of 4 percent exchangeable K in the cation exchange complex to avoid yield and/or quality loss of several fast-growing vegetables. Some problems associated with inadequate K under conservation tillage can be corrected by banding the K to reduce its fixation and obtain near-ideal absorption of the element. But as with P, the need for banding K appears to decrease with increased amounts of soil-test K.

Calcium and Magnesium

The lower pH associated with surface soils under conservation tillage leads to greater losses of Ca and Mg by leaching. The wetter conditions extend the period of leaching loss to the summer months, whereas losses for CT will usually be confined primarily to the winter and early spring.

The losses of Ca and Mg are closely associated with lower pH values, which can be corrected by liming. The choice of high-calcium liming materials or dolomitic materials containing more Mg should be made on basis of soil test results.

Micronutrients

Relatively little is known about the comparative availability of the micronutrients with conservation tillage versus CT. The availability of the micronutrients (B, Fe, Mn, and Zn) is probably favored by conservation tillage because the increased moisture tends to keep these elements soluble and aid in their diffusion and uptake. Chelation of Fe and Zn, which is favored with the additional organic materials, should also reduce fixation of these elements and, therefore, increase their availability. Humic substances that result from the decomposition of organic matter, and humic acid in particular, form complexes with micronutrients. Humic acid has been shown to improve Zn bioavailability more than HEDTA, a well-known synthetic chelating agent, while maintaining sufficient levels of other micronutrients for wheat grown in solution cultures (Mackowiak et al. 2001). On the other hand, Cu and Mo are probably less available with conservation tillage than with CT. The lower availability of Mo is expected because this element usually is less available under acid conditions, which are more prevalent with conservation tillage. A lower availability of Cu is expected because this element tends to become less available in soils of higher organic matter content.

PESTS

There are differences in pest problems and pest management between CT and conservation tillage. Conservation tillage, by greatly altering cultivation practices and allowing an accumulation of OM at the surface, presents an entirely changed environment for pest development. The change can have beneficial, negative, or neutral effects on pest development. It is difficult to predict whether conservation tillage will bring fewer pest problems than CT, but in all probability the pest problems will be different in the two production systems.

The actual problems appear to vary with different crops, climates, soils, pests, the extent of cover crop use, the manner in which residues of crops and cover crops are handled, and the number of years in conservation tillage. Despite these variables and the differences in problems, practical experience indicates less need for pesticide use with conservation tillage, although this saving in pesticides appears to be modified by cultural practices.

There is considerable variation in the amount of insect damage associated with conservation systems, and the relative damage between conservation and CT systems. Relative damage obviously will vary with different pests and crops. But other factors, such as length of time under conservation tillage, whether rotations are used, the amount of cover crops and/or compost used in the system, and the spray programs used also have an effect on the extent of insect damage. These variables, which have an important bearing on insect damage, appear to be important because of their influence on biological diversity and the effectiveness of beneficial organisms.

Conservation tillage lends itself to the use of crop rotations, which can have a very definite bearing on damage by several harmful insects.

The numbers of harmful insects are limited to some extent by the additional biological diversity associated with conservation tillage. Although crop residues allow many insect pests to overwinter, they are also a means of surviving the winter for a number of parasites or predators of these insect pests. Mulch and increased SOM in the upper few inches of soil favor a highly diversified biological population that limits development of any one species. Much of this is due to competition limiting food supplies, but the predatory or parasitoid activities are important aspects of limiting uncontrolled development of many harmful insects.

Beneficials

The value of beneficials for controlling insects was first brought to the attention of the senior author about 15 years ago by the virologist consultant Dr. Jack Simons, who was able to control several aphid-transmitted virus problems of cucurbits grown in Central America and the Dominican Republic by greatly reducing the use of broad-spectrum insecticides to control aphids. In their place, he substituted

Stylet-Oil, which tends to limit egg laying by aphids and greatly reduces their numbers, and limited the use of broad spectrum insecticides to suppress only serious outbreaks. The limited use of strong insecticides in conjunction with the reduced population of aphids kept viral infections at very low levels.

Much of the benefits of conservation tillage appear to be associated with the presence and maintenance of beneficial insects and mites. Some cover crops are better than others at supporting needed beneficials, and the degree of success may depend on whether beneficial insects are available to handle the problem and how well conservation tillage is able to support them.

The presence of beneficial insects and mites greatly increases as conservation tillage is practiced for a number of years, particularly if cover crops are fully utilized. Several insect pests kept under control with extra cover crops are described in Chapter 5. To keep beneficials effective, it is necessary to have adequate feeding areas present at all times during their active cycle. It perhaps is no coincidence that the greatest damage from whiteflies on melons grown in Guatemala occurs in the winter and early spring months, when much of the native vegetation, which supports beneficials, has withered due to extreme drought.

In northern climates, the surface layer of OM and SOM, increasing with time, tends to allow a number of beneficials to survive the winter, but their presence in sufficient numbers during the growing season often depends on the existence of sufficient native plants or cover crops to carry them through periods with insufficient pest numbers.

Insecticides and Beneficials

The effectiveness of beneficials can also be enhanced by proper use of insecticides that are chosen to limit invaders. Some insecticides can be devastating to beneficials. For example, most carbamates, organophosphates, and pyrethroids can literally wipe out predatory mites and cause pest mite outbreaks when applied to control other pests. However, the western predatory mite *(Metaseiulus occidentalis)* is resistant to sulfur and some carbamates and organophosphate, and can be used in treated orchards where it is effective against several plant-damaging mites (Flint and Dreistadt 1998).

In many cases, maximum effectiveness will be obtained if no insecticides are used, but often the numbers of invaders may be so great that the beneficials are overwhelmed and serious crop damage may occur. In such cases, insecticides will be needed, but harm to the beneficials can be limited by proper choice of the insecticide.

Some insecticides and acaricides that are highly toxic to natural enemies are acephate (Orthene), azinophosmethyl (Guthion), bendiocarb (Dycarb, Ficam), bifenthrin (Capture, Talstar), carbaryl (Sevin, XLR Plus), carbofuran (Furadan), chlorpyrifos (Dursban, Lorsban), cyfluthrin (Baythroid, Tempo), cypermethrin (Ammo), diazinon, dichlorvos (DDVP, Vapona), dicofol (Kelthane), dimethoate (Cygon), fenpropathrin (Tame), fenitrothion (Danitol), formetanate hydrochloride (Carzol), malathion, methidathion (Supracide), methomyl (Lannate), naled (Dibrom), oxamyl (Vydate), permethrin (Ambush, Pounce, Pramex), pyrethrin + piperonyl butoxide (Pyrenone), resmethrin, sulprofos (Bolstar), and sulfotep (Plantfume103) (Flint and Dreistadt 1998).

A number of insecticides that have little or no harmful impact on beneficials are *Bacillus thuringiensis* (Dipel), benzoximat (Banzomate), buprofezin (Applaud), clofentezine (Appolo, Acarastop), dicofol (Kelthane), diflubenzuron (Dimilin), fenbutatin oxide (Vendex), hexythiazox pirimcarb, piperonyl butoxide (Pybuthrin), pymetrozine, pyrethrin, pyriproxyfen, and tetradifon (Tedion V-18) (Heinz 2001).

Other insecticides or mixtures of insecticides can be used but may be limited to certain methods of application and certain beneficials. For example, imidacloprid can be used to control aphids and whiteflies if applied systemically and the beneficial wasp *Encarsia formosa* is used since it does not feed on plant sap. Imidacloprid can be detrimental to bumblebees and ought not be used for crops normally pollinated by bumblebees unless the crop is manually pollinated, as for greenhouse tomatoes. A mixture of fenbutatin oxide and hexythiazox can be used to control two-spotted mites without harming beneficials. Poor biological control of leaf miners can be corrected by a single spray application of cyromazine at 75 ppm, which stops the larvae of several leaf miner species without damage to the beneficials. Tebufenpyrad at 100 ppm active ingredient can be used to control *T. urticae* without damaging the parasitic wasp *E. formosa* (Heinz 2001).

Generally, insecticides specifically targeted for the pest rather than broad-spectrum ones will do less damage to the beneficials. Also, insecticides that break down quickly are preferred to ones that can remain active for long periods. Releasing beneficals after the temporary insecticide has largely dissipated helps ensure their full activity. For example, releasing *Orius* spp. or predatory mites to control western flower thrips about a week after applying dichlorvos, an insecticide with a short life, helps reduce harmful effects to the beneficials. The reduced numbers of western flower thrips increase the chances that the beneficials can keep this difficult pest under control. Applying the short-lived insecticide at times when the beneficial is most tolerant increases the chances for success. Generally, the pupae of beneficials are more tolerant to insecticides than the eggs or larvae. Also, spot treatment or treatment only of the heavily infested areas with insecticide can help in maintaining an effective population of beneficials. The localized treatments with insecticides can be aided by confining treatment to areas marked by geographic positioning systems (Heinz 2001).

Insect Control with GM Crops

Another approach for insect control, useful for both CT and conservation tillage, is the use of genetically modified (GM) crops that express an insecticidal protein from a soil bacterium, *Bacillus thuringiensis* (Bt). The protein is deadly to several lepidoptera insects. The primary insects controlled are European corn borer (*Ostrinia nubilalis* Hubner) and southwestern corn borer (*Diatrea grandiosella* Dyer). A new protein derived from Bt also provides protection against black cutworm *(Agrostis ipsilon)* and fall armyworm *(Spodoptera fragiperda)* and possibly western bean cutworm (*Richia albicosta* Smith), sugarcane borer (*Diatraea saccharalis* Fabricius), and lesser cornstalk borer (*Elasmopalpus lignosellus* Zellaer) (Babcock and Bing 2001).

PLANT DISEASES

Plants can be damaged or killed by a number of disease-causing microorganisms, including certain species of fungi, bacteria, viruses, and phytoplasmas. Diseases caused by bacteria and fungi are generally increased by the presence of moisture, suggesting that conservation tillage might be more responsible for disease outbreaks because of the additional moisture in the upper soil. Certain diseases also may be favored by cooler temperatures. The extra soil moisture and lower temperatures should be more of a problem with diseases that damage seeds and seedlings of spring plantings in northern climates. The slow emergence and early growth of many plants under such conditions predispose the plant to a number of diseases.

In addition to increased disease susceptibility, plant residues increased by conservation tillage may add to the problem by being a habitat for disease organisms after the affected crop is terminated. A number of plant pathogens overwinter in plant residues, including those which cause the following diseases: southern corn blight (*Helminthosporium maydis* Nisikdo), northern leaf blight of corn (*Helminthosporium turicum* Pass), corn anthracnose [*Colletotrichum graminicola* (Ces) C. W. Wils], yellow leaf blight of corn (*Phyllosticta maydis* Arny and Nelson), eyespot of corn (*Kabatiella zeae* Narita & Hiratsuka), tanspot of wheat [*Pyrenophora trichostoma* (Fr.) Fckl], Holcus leaf spot of corn *(Pseudomonas syringae)*, Goss's bacterial wilt and blight of corn (*Corynebacterium michiganense* ssp. *nebraskense*) (Boosalis et al. 1986).

Despite the possibilities of increased disease, actual damage caused by disease has been variable. There have been no serious outbreaks of these diseases with conservation tillage systems on the American Great Plains. The lack of serious outbreaks has been attributed to the weather, which tends to be hot, dry, and windy, and the relatively short period of time in which conservation practices have been used (Boosalis et al. 1986).

Actually, the additions of OM can have a wide range of effects on disease, particularly those that are soilborne. Sometimes the diseases are enhanced, but more often the added OM decreases the disease. Although some pathogens use the additional OM as an energy source to increase their inoculum, there are many more cases of microbial stimulation by the OM that can be harmful to the disease organisms. The effects are closely related to the OM added and its stage of decomposition. In intermediate stages of decomposition, some of the decomposition products can increase or decrease disease. The result depends to a large degree on the kind of pathogen and whether the products support the disease organism or favor other soil organisms that can more effectively compete for available food or are harmful to the disease organism.

Other factors limit disease development with conservation tillage. Attention was called earlier to the limiting of damage caused by *Pythium* spp., *Rhizoctonia solani,* and *Sclerotium rolfsii* as soil SOM levels increased, probably due to better water infiltration and improved air exchange in the soil. It may take several years of conservation tillage to appreciably suppress the development of these pathogens if there is a high level of disease inoculum.

Benefits from Reduced Spraying

Additional benefits from cover crops and reduced tillage also appear to be present. Incorporating plant debris to limit disease losses can interfere with the maintenance of insect habitat and decreases weed control benefits and, therefore, may not be the most desirable approach for control. At times, better control of disease can be obtained by the use of cover crops that limit the need for cultivation and spraying of the primary crop. The reduction in spray applications and cultivation can limit damage to the protective waxy leaf surface layer that is the first physical barrier to disease invasion. By limiting spraying, the application of pesticides, soaps, surfactants, spreaders, and stickers is reduced, and greater numbers of beneficial microorganisms capable of competing with the disease organisms are maintained on the plant's surface (Phatak 1998).

Rotations and Disease Control

Disease can at times be controlled by rotation of crops, an option that is easily taken advantage of with conservation tillage. Rotation of crops, which is so essential for good conservation practices, helps to maintain a diverse population of soil microorganisms that tend to prohibit dominance by any single organism. By rotating crops, a number of soil-inhabiting pests are denied the essentials for rapid buildup, since many of these pests depend on a single crop or related crops for their maintenance. A number of plant diseases limited by rotating crops are presented in Table 10.2.

TABLE 10.2. Some plant diseases controlled by crop rotation

Disease	Crop	Causal agent	Year on	Year off	Safe crops
Cereal scab	Barley, corn, rye, wheat	*Gibberella saubinetti* (Mont.) Sacc.	1	1-2	Alfalfa, oats, clover, flax, soybeans
Flag smut	Wheat	*Urocystis tritici* Koern	1	1	Any but susceptible wheat
Diplodia	Corn	*Diplodia zeae* (Schw.) Lev	1	1-2	Any but susceptible corn
Wilt	Alfalfa	*Phytomonas insidiosa* (LMcC.) Bergey et al.	3-4	–	Any crop including resistant alfalfa
Black rot	Sweet potatoes	*Ceratostomel-la fimbriata* (Ell Haas)	1	3-4	Seed and seedbed free of parasite
Foot rot	Sweet potatoes	*Phenodomusde-struens* Hart	1	2-4	Seed and seedbed free of parasite
Scurf	Sweet potatoes	*Moniilochaetes infuscans* Hals	1	3-4	Seed and seedbed free of parasite
Anthracnose	Beans	*Colletotrichum lindemuthianum* (S. & M.)	1	2	Any crop but beans
Blight	Beans	*B. phaesoli* E. F. S., *B. medicaginis* Link & Hall	1	2	Any crop but beans
Wilts	Potato, tomato, eggplant	*Verticillium* spp., *Fusarium* spp.	1	3-5	Small grains, grass, corn, legumes
Bacterial canker	Tomato	*Aplanobacter michiganese* E. F. S.	1	2	Any crop but tomato
Anthracnose	Cucumber	*C. lagenarium* (Pass.) E. & H.	1	2	Any crop but cucurbits
Bulb rot	Onions	*Fusarium* spp.	–	–	Any crop but onions
Texas root rot	Cotton	*Phymatotrichum omnivorum* (Shear)	2-3	3-4	Small grains, sorghum, corn, grasses
Black shank	Tobacco	*P. parasitica nicotianae*	1	4	Any crop but tobacco
Black root rot	Tobacco	*Thelaviopsis basicola* Zopf	1-2	5-6	Grasses
Granville rot	Tobacco	*B. solanacearum* E. F. S.	1	3-4	Corn, small grains, grasses, cotton, sweet potatoes

Disease	Crop	Causal agent	Year on	Year off	Safe crops
Damping off	Sugar beets	*Pythium* spp., *Rhizoctonia* spp.	1	4-5	Most cultivated crops
Leaf spot	Sugar beets	*Cercospora beticola* Sace	1	4-5	Most cultivated crops

Source: Leighty, C. E. 1938. Crop rotation. In *Soils and Men: USDA Yearbook of Agriculture* (pp. 406-430). Washington, DC: USDA.

Some rotations, especially those that include safflower, favor the development of mycorrhizae. As indicated in Chapter 5, the infection confers certain advantages to the affected plants, one of the major benefits being the increased resistance to certain fungi (*Fusarium, Phytopthera,* and *Pythium*).

Healthier Plants and Disease Control

Conservation tillage also can have indirect beneficial effects on plant diseases. Probably through control of erosion and the maintenance of better moisture and nutrients, it tends to produce healthier plants that are more resistant to disease.

As indicated in Chapter 5, some disease organisms are consumed by certain amoebas and nematodes, and some microbes are parasitized by microorganisms, but specifics of control are lacking, making it difficult to put them to practical use. On the other hand, our knowledge of antibiotics has advanced sufficiently to utilize this mode of action to control several plant diseases. Streptomycin, a product formed by the soil bacterium *Streptomyces griseus,* is being used as a seed dip against damping off and as a cutting dip or foliar spray to control bacterial leaf and stem or wilt diseases. Various members of *Bacillus, Penicillium,* and *Pseudomonas* spp. are applied as root dips or seed inoculants to control damping-off caused by *Phytophthora, Pythium,* and *Rhizoctonia* (Flint and Dreistadt 1998).

Fungicides and Bactericides

Despite some benefits in suppressing soilborne diseases and helping to maintain plant defenses by limiting damage from cultivation and some pesticides, it will be necessary to use fungicides and/or bactericides with conservation tillage to keep certain diseases in check. Such use is almost a foregone conclusion if warm temperatures and elevated humidity favorable to the disease organisms are encountered.

Fungicide Effects on Beneficials

It is important that these protective agents do not unduly harm beneficial organisms. As indicated for insecticides, less harm will be done if the application of fungicides is limited to times when the beneficials are most tolerant of them.

Some fungicides that can be used without harm are anilazine, bitertanol, bupirimste, carbendazim, captafol, captan, chlorothalonil, copper-oxychloride, ditalimfos, dithianon, flutriafol, hexaconazole, iprodione, mancozeb, procymidone, thiram, thiophanate-methyl, triadimefon, triadimenol, tridemorph, triforine, and vinclozolin.

NEMATODES

The role of conservation tillage in suppressing nematodes appears to be ambiguous. There are reports of lessened nematode populations with higher SOM, especially after certain rotations (Table 9.2). Some of this may be related to the controlled growth of harmful nematodes due to increased competition from numerous other organisms or due to actual predation by parasitic nematodes or other soil organisms.

There is a general consensus that if nematodes are not a problem, conservation tillage generally will keep damaging nematodes from becoming a problem even if susceptible crops are grown. The chances of trouble are further reduced if only seeds or plants free of nematodes are used, crops are rotated with nonsusceptible crops, and if certain crops with nematicidal effects are grown.

The grasses and small grains are usually suitable to grow with crops susceptible to root-knot nematodes. Alfalfa, forage grasses, velvet beans, and sorghum, as well as grasses and small grains, can be grown in rotation with cotton, peanuts, or tobacco to limit root-knot damage.

Cover crops can be grown to limit nematode development. Crotalaria can serve as a trap plant for root-knot nematodes. Although the nematodes can attack crotalaria, they cannot reproduce in the plant, limiting buildup. Other cover crops that tend to limit development of nematodes are sorghum-sudangrass hybrids, hairy indigo, showy crotalaria, sunn hemp, velvet beans, mustard, and marigolds (Phatak 1998).

Rapeseed (canola) tends to be slightly more effective in suppressing root-knot nematodes than sorghum-sudangrass hybrids. Sorghum-sudangrass hybrids and sudangrass have shown considerable but not consistent suppression of root-knot nematodes. Some of the inconsistencies may be due to the variation in effectiveness of different cultivars. Control also depends on timing and incorporation of the cover crop, as is the case with cut or chopped sudangrass, which needs to be incorporated before frost lest the nematicidal effect be lost (Clark 1998).

Not all pathogenic nematodes are controlled by specific cover or rotational crops. Rather, most plants that are effective suppress only certain nematode species, while other plants may not be affected or at times may be used by nematodes as plant hosts.

A number of natural enemies limit nematode numbers and destructiveness. In addition to nematodes feeding on other nematodes, amoebas, flatworms, mites, nematode-trapping fungi, protozoa, and springtails limit nematode damage to crop plants. Since conservation tillage tends to increase the numbers and species of these nematode enemies, we can expect some benefits from reduced tillage. Reduced tillage appears to be especially beneficial to predaceous nematodes, as they usually are found in substantial numbers in perennial crops and appear to be adversely affected by cultivation (Flint and Dreistadt 1998). Conservation tillage, especially no-till, might have some advantage in nematode control because of limited disturbance of predaceous nematodes.

Although several of these natural enemies in all probability limit pathogenic nematode damage, there has been little practical application of their use. Nematodes that feed on nematodes are available for control, but the authors do not know of bacteria or fungi being offered for sale that are capable of destroying plant pathogenic nematodes.

WEEDS

Reduced tillage often requires other strategies for weed control, since weed control, in CT often is provided by cultivation. Emphasis may need to be placed on herbicides, and use of GM crops that have tolerance to herbicides are effective, but there are other strategies, such as smothering with cover crops or mulch to control weeds in conservation tillage.

Not only are methods of weed control somewhat different for conservation tillage than for CT, but weed problems may be different. Deep tillage tends to bury some seeds so that they do not readily germinate unless they are brought close to the surface by subsequent plowing or disking. Conservation tillage, especially no-till, tends to keep weed seeds close to the surface, which is satisfactory for germinating small-seeded broadleaf weeds and annual grasses, but not suitable for germination of the large-seeded weeds. Because many of the large seeds will not germinate with no-till, their numbers will usually be reduced with time, making weed control appreciably different after some years in the system. Whereas simple perennials (weeds that produce a shoot each year from a main taproot) seldom become a problem under CT, both simple and creeping perennials can become difficult to control under NT.

Weed control in conservation tillage as well as in CT is aided by (1) thoroughly composting animal manure to eliminate weed seeds, (2) keeping weeds from going to seed, (3) power washing tillage equipment after using in fields with weed problems, and (4) keeping alleys and field edges mowed to avoid weed seed production (Grubinger 1997).

Since cultivation is limited in conservation tillage, weeds are largely controlled by biological means and herbicides. At first, more herbicides may be needed with conservation tillage, but in time as more SOM is built up, there should be no greater need for herbicides with conservation tillage than with CT. After a while, there may be less need for herbicides with conservation tillage, especially if sufficient efficient cover crops are used and if they are handled wisely.

Biological Weed Control

Biological control of weeds depends on competition with other plants, smothering effects of residues or mulches, allelopathic effects of residues, and damage done to weeds by herbivores, pathogenic microorganisms, beneficial insects, and mites. These several means of controlling weeds have been touched upon in Chapters 5, 6, and 8, but a more detailed presentation of these controls is given in this section.

Rotations

Rotations can help control weeds, although complete control by rotations is impossible. What we can expect of rotations is that they introduce other plants that can compete more effectively with the weeds by crowding, or they add certain practices, such as mowing, that will limit the production of weed seeds. Alfalfa is notable for lessening the damage from noxious weeds. The vigorous growth of alfalfa tends to limit vegetative weed growth, and the frequent cuttings tend to limit the production of viable weed seeds.

Mulches

Practical weed control is possible with mulches, but the extent of control may vary depending on the type of mulching material, its depth, weed species, environmental conditions, and type of primary crop. Often, some herbicides in addition to the mulch may be needed to obtain effective control, especially if large numbers of noxious weeds are present or in very humid climates, where a number of weeds can get established on the mulch. The need for herbicide is reduced if the primary crop can quickly produce a canopy that can shade the weeds.

Dead mulch of materials such as hay, straw, pine needles, wood chips, recycled paper pulp, or composts at depths of 10-15 cm (4-6 inches) has been used to discourage weed growth. Immature compost appears to have greater effectiveness than well-cured material, possibly due to allelochemicals, which may also have some adverse effects on germination and seedling growth of the primary crop (Stoffella et al. 1997). Utilizing well-cured composts can avoid the inhibitory effects of mulch on the primary crop, but it appears that the mulch needs to be maintained at greater depths, which may require some additions as the season progresses. Dead mulch can be expected to be more effective in drier climates.

As discussed in Chapter 5, living mulches also can be used for weed control, but competition from the mulch can reduce crop yields. Michael Costello of the University of California has developed a living mulch system that allows broccoli to be grown without herbicides. Early attempts failed because the site was a clay loam that restricted root growth because of poor soil porosity. By moving the experiment to a well-drained sandy loam soil, mowing of the living cover early, and supplying extra water and nutrients, he was able to produce broccoli yields similar to those of clean cultivation (Grossman, 2001).

Cover Crops

Cover crops can have considerable influence on the weed problem, but weed control varies with the kind of cover crop and how it is handled. Oats, rye, and buckwheat among the nonlegumes, and the legumes berseem clover, cowpeas, subterranean clover, and woolypod vetch have been listed as outstanding weed fighters. A list of effective cover crops for suppressing weeds in different regions of the United States was given in Table 5.1.

Cover crops used for weed control are killed by undercutting, flailing, or with herbicides. The dead cover is left on the surface as mulch, where it also benefits the crop by providing better soil moisture while reducing erosion and lessening soil temperature.

The killed mulch approach usually allows several weeks of essentially weed-free conditions. In that time, the planted crop has an excellent chance to get well established without serious weed competition, allowing it to make a normal crop without additional weed control.

A study of seven clovers in Canada for controlling mustard showed that control depended upon the weed populations, with greater suppression of mustard with clovers on a low-productivity site than on a high-productivity site. Alsike and berseem suppressed mustard in both years of the study, whereas balansa and crimson were beneficial in only one of the years. The annual clovers (balansa, berseem, crimson, and Persian) failed to show any advantages over the perennials (alsike, red, and white clovers) (Ross et al. 2001). Large-seeded legumes and sweet clover gave better weed control under dryland prairie conditions than *Trifolium* spp. clovers (Schlegel and Havlin 1997).

The value of the cover crop as a weed suppressor depends on several factors, such as its allelopathic qualities, its ability to start quickly and compete with weeds, and the biomass produced.

Allelopathic qualities. Allelopathic substances can act as a double-edged sword. While the allelopathic qualities of some cover crops can suppress weeds, these same qualities may also hinder the primary crop. In selecting cover crops, it is desirable to know the impact of the cover crop on the primary crop. Fortunately, some of these allelopathic properties dissipate quickly, providing some early weed suppression and diminishing sufficiently by the time the crop emerges to have little effect on the primary crop. The primary crop is aided by the fact that it is planted at a depth where it may not be in contact with the cover crop and so not directly affected by the allelopathic substances for some time. Moving residues to either side of the plant row just prior to planting can provide an allelopathic-free area so that the primary crop can develop normally.

The handbook *Managing Cover Crops Profitably* (Clark 1998) lists rye and sorghum-sudangrass hybrids as excellent, and barley, oats, buckwheat, and subterranean clover as very good for suppressing weeds because of the cover crop's allelopathic qualities. Other cover crops listed are black oats, hairy vetch, sweet clover, and woolypod vetch.

Other workers also have listed the outstanding allelochemical effects of rye in suppressing weeds and have cited legumes as being effective, including crimson clover, hairy vetch, and subterranean clover. A living mulch of subterranean clover supplied nearly complete weed control in corn (Worsham 1990).

Early competition. Generally, small-seeded clovers establish rather slowly and may be at a disadvantage as compared to the larger-seeded clovers or grasses in suppressing weeds. The disadvantage appears to be greater if the weed problem is severe. The annual clovers (crimson, berseem, and Persian) tend to be more upright and have an early advantage over some perennial clovers. Subterranean clovers, although annuals, tend to have a prostrate growth but are excellent weed suppressors because of their rapid buildup of biomass.

Biomass. Weed suppression from mulch is often due to its smothering effect. By limiting light and physically interfering with weed seed emergence, large amounts of biomass effectively block competing weeds, giving the primary crop with its small area free of mulch a competitive edge in the battle for light and nutrients. Aside from any allelopathic effect, this smothering effect primarily depends on the amount of biomass produced, which is dependent on such factors as the type of cover crop or residue, soil fertility, climatic conditions, and age of cover crop when terminated. Some typical biomass amounts produced by several cover crops are listed in Table 10.3.

Table 10.3 indicates considerable variation in biomass produced by different cover crops, which is much greater for some nonlegumes than most of the legumes. The smaller variation in yield of legumes could well be due to the nitrogen fixation qualities of the legumes. The senior author has found that the yields of late-seeded non-leguminous cover crops (rye and annual ryegrass) grown on highly

TABLE 10.3. Amounts of biomass produced by several cover crops

Cover crop	Dry matter produced (lb/acre)
Legumes	
Berseem clover	6,000-10,000
Cowpeas	2,500-4,500
Crimson clover	3,500-5,500
Field peas	4,000-5,000
Hairy vetch	2,300-5,000
Medic	1,500-4,000
Red clover	2,000-5,000
Subterranean clover	3,000-8,500
Sweet clovers	3,000-5,000
White clover	3,000-6,000
Woolypod vetch	4,000-8,000
Nonlegumes	
Annual ryegrass	2,000-9,000
Barley	3,000-10,000
Buckwheat	2,000-3,000
Oats	2,000-10,000
Rye	3,000-10,000
Wheat	3,000-7,000
Sorghum-sudangrass	8,000-10,000

Source: Clark, A. (Coordinator). 1998. *Managing Cover Crops Profitably,* Second Edition. Handbook Series Book 3. Beltsville, MD: Sustainable Agriculture Network, National Agricultural Library.

leachable sandy loams in southern New Jersey often were more than doubled by a topdressing of N, although mixed fertilizer with N gave slightly greater increases.

Some of the variation in biomass yields of individual cover crops could well be due to factors other than fertility. Climatic conditions, planting rates, length of growing period, and pest damage could influence the biomass produced.

Obviously, large amounts of biomass are essential for good weed control by cover crops, and the mulch needs to be evenly distributed to eliminate or reduce application of postemergence herbicides. Three or more tons per acre of grass (cereal grains) or mixtures of grass with legumes have been needed to suppress weeds in no-till pumpkins. Seeding rates of grass crops in mixtures can be reduced by 50 percent and those of legume crops by 25 percent (Morse et al. 2001).

Rye, which is adapted to most areas of the continental United States as a cool-season cover crop, is noted for the biomass produced. It is relatively cheap and easy to establish by overseeding, where it will make little growth until cool weather sets in, thereby providing little competition with the primary crop. Partly because of its fast growth in cool weather, it is outstanding as a suppressor of chickweed, foxtail, lamb's quarter, redroot pigweed, and velvetleaf. Rye also has allelopathic qualities that help suppress dandelions and Canada thistle, as well as some thiazine-resistant weeds (Clark 1998).

Sorgum-sudangrass hybrids are also noted for the amounts of biomass produced but because of their dependence on warm weather are limited to summer plantings between crops in most of the United States. They can be grown year-round in much of Florida, small portions of southern Texas and Louisiana, and the tropics.

In addition to their allelopathic control of several weeds (barnyard grass, crabgrass, green foxtail, redroot pigweed, smooth pigweed, common ragweed, purslane, and velvetleaf), the sorghum-sudangrass hybrids also tend to smother weeds. The smothering effect is enhanced by higher seeding rates than for forage plantings (50 lb/acre for broadcast or 35-40 lb/acre for drill). Biomass and potentially increased smothering effects, enhanced on fertile soils that receive ample water, can also be increased by multiple cuttings (Clark 1998).

Greater biomass often can be obtained by combining two or more cover crops. Several efficient cover crop combinations are presented in Chapter 6. From the standpoint of biomass, combinations using a climbing legume with a tall upright grass can be highly productive. One of the more efficient mixtures for producing considerable biomass in the Northeast has been a mixture of rye and vetch, grown during cool weather. For warm-weather growth, sorghum-sudangrass hybrids can be combined with buckwheat, cowpeas, forage soybeans, or sunn hemp. In these mixtures, buckwheat gets an early start in suppressing early weeds. The sorghum-sudangrass helps support the sprawling sesbania, forage soybeans, and cowpeas. Sesbania, cowpeas, and sunn hemp are legumes and can supply needed N for the sorgum-sudangrass to provide larger amounts of biomass.

Weed Control by Vertebrates, Invertebrates, and Pathogenic Organisms

Worldwide biological control of weeds by vertebrates, invertebrates, fungi, and native organisms has been catalogued. The catalog lists a wide variety of agents used for weed control, the target weeds, location of trials, degree of success in controlling weeds, and the survival extent of introduced organisms (Julien 1992).

Biological control of weeds by herbivores and pathogenic organisms in the United States has largely been confined to noncropped areas of woodlands, rangelands, and roadsides. This may be changing as there is more emphasis on reducing use of herbicides, although the increased use of cover crops may take up much of this slack.

Herbivores, both vertebrates and invertebrates, take their toll of weeds. Conservation tillage allows more grazing by large vertebrates (cattle and sheep), and it also increases the numbers of small vertebrates (birds, rodents, prairie dogs, moles) that feed on seeds, roots, and leaves of many weeds. The damage to weeds by insects and mites may exceed the damage caused by the vertebrates. Insect introduction has helped curb a number of damaging weeds (Klamath weed or St. John's wort, puncture vine, Mediterranean sage, Scotch thistle, tansy ragwort, alligator weed, and musk [nodding thistle]).

Two pathogenic fungi, *Colletotrichum gloesporiodes* ssp. *aeschynomene* and *Phytophthora palmivora*, have been found to have weed control properties. These organisms may be collected and offered for sale, but it is advised that local departments of agriculture be contacted before purchasing them (Flint and Dreistadt 1998).

Weed control by invertebrates and pathogenic organisms is presented in Table 10.4.

TABLE 10.4. Invertebrates and fungi capable of controlling weeds*

Weeds		Biological control agents	
Common name	Scientific name	Common name	Scientific name
Canadian thistle	Cirsium arvense	Stem-mining weevil	Ceutorhynchus litura
		Seed weevil	Larinus planus
		Gall fly	Urophora cardui
Diffuse knapweed	Centaurea diffusa	Seedhead gall fly	Urophora affinis
Spotted knapweed	Centaurea maculosa	Seedhead gall fly	U. quadrifasciata
		Root-boring weevil	Sphenoptera jugoslavica
Gorse	Ulex europaeus	Seed weevil	Exapian ulicis
Italian thistle	Carduus pycnocephalus	Seedhead weevil	Rhinocyllus conicus
Milk thistle	Silybum marianum		
Klamath weed	Hypericum	Root-boring beetle	Agrilus hyperici
or St. John's wort	perforatum	Leaf beetle	Chrysolina hyperici
		Leaf beetle	C. quadrigemina
		Gall midge	Zeuxidiplosis giardi
Leafy splurge	Euphorbia esula	Stem and root borer	Oberea erythrocephala
Mediterranean sage	Salvia aethiopis	Leaf-, root-chewing, and mining weevil	Phrydiuchus tau
Milk thistle	Carduus nutans	Seedhead weevil	Rhinocyllus conicus
or nodding thistle		Crown and root-mining weevil	Trichosirocalus horridus
Northern jointvetch	Aeschynomene virginica	Collego**	Colletotrichum gloeosporioides ssp.
Puncture vine	Tribulus terrestris	Seed-eating weevil	Microlarinus lareynii
		Crown- and stem-mining weevil	M. lypriformis
Scotch broom	Cytisus scoparius	Twig-mining weevil	Leucoptera spartifoliella
		Seed-eating weevil	Apion fuscirostre
Spurge	Euphorbia spp.	Leaf-, root-feeding flea beetles	Aphthona cyparissiae
			Aphthona nigriscutis
		Gall midge	Spurgia esulae
Stranglevine	Morrenia odorata	DeVine**	Phytophthora palmivora
or milkweed vine			
Tansy ragwort	Senecio jacobaea	Crown-, leaf-, stem-feeding flea beetles,	Longitarsus jacobaeae
		Shoot-feeding cinnabar moth	Tyria jacobaeae
Tumbleweed	Salsola australis	Leaf-mining and chewing moth	Colephora klimeschiella
or Russian thistle	(= S. iberica)	Stem-boring moth	C. parthenica

TABLE 10.4 *(continued)*

Weeds		Biological control agents	
Common name	Scientific name	Common name	Scientific name
Yellow starthistle	*Centaurea solstitalis*	Seedhead weevil	*Bangasternus orientalis*
		Seedhead gall fly	*Urophora sirunaseva*

Source: Flint, M. L. and S. H. Dreistadt. 1998. *Natural Enemies Handbook.* University of California, Division of Agriculture and Natural Resources Publication 3386. Berkeley, CA: University of California Press.
*Most of the insects have been introduced for classical biological control. Some are quite effective naturally in some areas but, if not present, may be purchased. Contact local county departments of agriculture before purchasing as there may be limitations as to their use.
**Commercial sources, but these may be uncertain.

Herbicides

Although mulch may be used to control weeds in certain cases, much of the weed control in conservation tillage, especially in the early years, will probably have to be done by herbicides. Early preplant applications (10 to 45 days before planting) can give excellent control of annual weeds. Using full season herbicides tends to give season-long weed control, although split applications of medium-longevity herbicides can also be used. Kill of active weeds will be increased by adding crop oil or surfactant to the herbicide, but if there is abundant growth of 1 to 3 inches winter or early summer annual weeds, use a burndown herbicide such as Roundup or Paraquat prior to planting or crop emergence.

The pre-emergence applications can consist of a full-season program that will control weeds for most of the growing season or a short residual program that provides control until the crop can shade out the weeds. The short residual program works well with crops that provide shade quickly as with plants in narrow rows.

Postemergence treatments can be used if an earlier preplant application is not used or if perennial weeds are a problem. Weeds need to be identified and recommended herbicides applied. Applications to control difficult perennials need to be timed properly so that the applied herbicide can be readily translocated to the root. The most effective times to control difficult perennials are prior to full bloom in the spring or as new growth appears in the fall (Childs et al. n.d.).

Certain herbicides can be applied postemergence to herbicide-tolerant crops without any damage to them. Genetically modified seeds of several major crops are now available that are tolerant or highly resistant to specific herbicides. Typical crops include Pursuit tolerant (imi) corn, STS soybeans, Roundup Ready soybeans, Poast compatable corn, and Liberty Link corn. The specific herbicide to which the GM crop is tolerant can be sprayed without damage to the crop.

Nonselective postemergence sprays on non-GM crops will have to be applied with shielded hoods or other means so that the primary crop does not receive any of the herbicide. Wipe-on applicators that reach only the weeds may be another approach. This approach works well with weeds that are taller than the primary crop, but spot treatments can be applied by hand on weeds of various heights.

Between-Row Mowing Plus Herbicide

Weeds can be controlled in soybeans by a combination of band-applied herbicide over the crop row and two between-row mowings. The mowings, when tallest weeds were 8 to 24 inches tall, have effectively controlled annual grass and broadleaf weeds, primarily giant foxtail, common ragweed, and waterhemp species, eliminating the need for herbicide beyond applied preplant. Weed control was enhanced by using varieties and row spacings that favored early canopy closing (Donald 2000).

MISCELLANEOUS PESTS

Birds

Fields under conservation tillage suffer occasionally from bird damage, primarily soon after harvesting small grains, when birds come to feed on spilled grain or on insects disturbed by the harvesting process. Damage to stands of emerging seedlings in no-till plantings soon after harvest can be severe. Some problems can be averted by careful control to avoid spilling of harvested grain and better coverage of planted seeds. Seeds treated with repellents are available but may be expensive. Noisemaking devices appear to have little impact and are expensive (All and Musick 1986).

Rodents

Field mice, prairie voles, and southern bog lemmings have caused damage in conservation-tillage crops. Field mice seldom cause serious economic damage, except in small fields within fencerows with dense vegetation next to farm buildings or feedlots and in newly planted orchards. Damage from southern bog lemmings is also infrequent but damage from prairie voles can be extensive in no-tillage corn.

Control has been limited to high hazard conditions of increasing populations or when rodents are in a reproductive phase. Suggested controls consist of plow-tillage, close cutting and removal of vegetation, or disking sod fields prior to planting (All and Musick 1986).

The senior author has witnessed control of field mice in young orchards by eliminating much of the vegetation close to the tree trunks and by using rodenticides.

Slugs

Slug damage can be severe in conservation-tillage crops, with most damage occurring in warm wet springs in crops with heavy mulch, considerable crop residues, large applications of manure. Control measures have been used for situations where large populations are expected. If slugs have been a problem in the past or a cool wet spring is anticipated, destruction of heavy mulch accumulations by some tillage or delaying planting until warmer weather are suggested means of control (All and Musik 1986).

Snails and slugs have natural enemies that help keep them under control. Some species of snails are predaceous, feeding on other snails. The decollate snail can keep brown garden snails in citrus groves under control, but their use is limited to certain San Joaquin Valley and southern counties in California (Flint and Dreistadt 1998).

Chapter 11

Putting It All Together: Combining Organic Matter Additions with Conservation Tillage

The combination of large OM additions with conservation tillage has proven to be a means of increasing or at least stabilizing SOM while maintaining agricultural production. The positive change in SOM with its accompanying beneficial effects on soil productivity promises a sustainable soil, which is essential for developing a sustainable agriculture.

The kind and amount of OM as well as the manner in which it is added have an important bearing on the degree of positive change. Probably equally important are cultural methods that combine reduced tillage (conservation tillage) with additions of OM, most of which is kept as mulch. The combination system can be conveniently designated as COM.

BASIC APPROACHES

There are essentially two different approaches to handling the combined system of increased OM additions with conservation tillage. One method practically abandons all so-called chemical inputs, limiting the application of fertilizers and pesticides to natural products. The second approach realizes that COM allows substantial reduction in inputs, but sustainable production (economically viable and in sufficient quantities as well as of good quality) is not possible without supplementing the addition of extra OM with suitable products to adequately control pests and supply nutrients in a timely fashion. These inputs can be synthetic or natural products.

The first method, commonly known as organic gardening or farming, depends on manure, legumes, composts, earthworm casts, animal and crop residues, and various organic wastes for N and other nutrients. Steamed bone meal and rock phosphate are used to supply P. Ground rocks and minerals are considered safe and so materials such as feldspar, glauconite (greensand), alunite, and illite are used primarily as K sources. Rocks, minerals, and glacial rock dust as well as manure, composts, and animal and plant residues are used as sources of micronutrients. Pests are controlled by horticultural oils, insecticidal soaps, diatomaceous earth, copper, sulfur, and some natural products such as rotenone, pyrethrum, Bt, neem extract, and ground seashells. These materials are often combined with the use of beneficials and various repellants or traps.

The second method does not discriminate as to whether the material is natural or synthetic. Rather, there are three primary concerns for using various products: (1) that products supply needed nutrients in a timely fashion and that pests are reasonably controlled; (2) the product must be economical to use so that the increase in yield or reduction in pest damage will more than cover costs; and (3) the products need to perform with limited damage to plants, animals, or humans, and their use over the years should not adversely affect soil health or cause pollution of ground or other sources of water.

In the past, organic farming has produced less per acre (about 5 to 10 percent) than systems using unlimited inputs. Quality as well as quantity was reduced, but organic farmers have been able to compete and increase in numbers in many instances because of greatly reduced input costs and higher prices received for organic products. Whether the system will be as profitable when nearly everyone is producing organic foods remains questionable. This evaluation may not be long in coming because of a greatly increased demand for organic foods. The increase in demand does not seem justified—at least in recent years—since food produced with intensive inputs appears to be as nutritious and healthy as the organic product, and often more appealing. Although such food may contain traces of pesticides, these are kept within acceptable limits that may be less harmful than some bacteria, fungi, or insects remaining on the organic food not receiving adequate pest control.

Methods Maximizing the Combination

The authors feel that neither organic farming or unlimited use of chemical inputs can produce maximum economic yields (MEY) capable of providing sustainability for increased populations. It is our belief that the system of limited inputs now being practiced by organic farmers probably is not capable of producing sufficient food and fiber even if all farmers adopted it. The system may fall even further behind as future population increases dramatically. The major system of unlimited inputs also does not seem to offer a sustainable agriculture that can take care of future generations. Several factors appear to be involved. Pollution of groundwaters from excessive use of pesticides and fertilizers and the depletion of natural resources are part of the problem, but the loss of SOM due to continuous tillage, limited OM additions, and erosion appears to degrade soils sufficiently to limit their productivity. By combining the best of both systems, there is a much better chance of reaching yield goals that can be sustained for many years.

The COM system offers a much better chance of supplying sufficient food and fiber for increased populations. Adding large amounts of OM and maintaining much of it on the surface provides many of the benefits associated with organic farming, allowing some reduction in inputs, which increases the chances for profitability while reducing the chances for pollution. The use of fertilizers and pesticides that can do the best job, regardless of whether they are organic or synthetic, can produce better crops with limited damage, thereby increasing the amount and quality of the products produced. For example, quick-acting chemical fertilizer, with its greater solubility, can rapidly supply needed nutrients (especially N) that may be released too slowly from organic sources to support the rapid growth needed soon after plants emerge. Likewise, depending primarily upon natural insecticides to control sudden large populations of insect pests can result in considerable loss of yield and quality. Occasional use of a quick-acting synthetic pesticide, specifically targeted for the invading pest, often can control the outbreak with minimal damage to the crop and still allow pest control by beneficials.

The COM system is being used in many areas although its use is relatively young (20 to 30 years), and is still undergoing changes to meet variable demands. While practical programs have been established, there is a great need to work out the ideal system for different regions and crops. For example, the best means of adding OM to the system for various localities, soils, and primary crops must be determined.

However, a great deal of information is available that enables us to reach interim decisions to make COM profitable now, while nurturing the system until such time that there may be more positive answers to pressing questions. The questions are primarily in the areas of (1) adding OM, (2) supplying needed nutrients, (3) controlling pests, (4) adding water, and (5) providing suitable machinery to do the job. Essentially, these same questions are pertinent in any farming system, but because we are making major changes in a soil's ability to produce a crop by using the new system, some old methods of handling the five primary questions may no longer be applicable.

While COM offers the best chance for sustainable agriculture, its success depends to a great extent on how it is used. A brief examination of applicable methods of answering the basic questions could be useful in making the COM system a viable approach to maintaining agricultural sustainability.

ADDING ORGANIC MATTER

The success of COM depends closely on adding sufficient OM. We contend that for a long time to come, sustainable agriculture is only possible if sufficient organic matter is present to maintain a diverse soil biological population and provide certain soil characteristics that help limit erosion and create a proper balance of air and water in the soil. As seen in previous chapters, the primary modes for adding OM are to grow it in place or bring it in from other sources. It may be helpful to summarize our current ideas about the various forms of OM available under the two methods and their relative usefulness.

Growing OM in Place

Growing OM in place is the cheapest and for many growers may be the only way of increasing the OM supply. Traditional animal production, where growers are still producing a large portion of animal feeds, allows for the economical growing of sods, pastures, and forages, all of which tend to add significant quantities of OM. This cannot apply for most animal factories that purchase a major share of animal feeds. Nor can it be economical for many farms that do not have substantial animal populations.

Cover Crops

For many growers, cover crops are the primary source of added OM. There are very few farms where cover crops cannot be used to good advantage. Cover crops can produce significant quantities of OM by utilizing periods between production of primary crops. Started while the primary crop is still growing but after most of the primary crop's growth has been made allows enough time to produce substantial quantities of OM from a number of cover crops that grow during the cool autumn period, overwinter, and produce a great deal more growth during the cool spring period. Other cover crops that grow best during warm periods are ideally suited for periods between primary crops in temperate or tropical climates. Considerable production of vegetables now takes place in tropical regions during the winter period for shipment to northern countries, leaving several idle summer months that are ideally suited for growing heat-tolerant cover crops.

As listed in Chapter 6 and in much greater detail in the book *Managing Cover Crops Profitably* (Clark 1998), a number of cover crops can be used in various regions. The choice may depend on such factors as the type of soil, climate, time available for growth, the primary crop grown, and whether there are special problems such as erosion or pests. In addition, a great deal of local information about cover crops is usually available from county extension agents and local farmers. A great deal more information will probably be available in the near future, particularly about new sources of cover crops, as considerable attention is being paid to this area of research.

It is important that the cover crop produce the maximum amount of OM that is compatible with the primary crop. For cover crops that overwinter, this often requires seeding the cover crop in the previous primary crop in its later stages of growth, and then allowing it to grow long enough in the spring. The early seeding and late termination usually produce extra OM. In many cases, the termination of the cover crop can be extended only if there is sufficient moisture, making irrigation of the cover crop helpful during dry periods.

Yields of the cover crop depend on sufficient nutrients. Often enough nutrients are left over from the primary crop to grow a substantial cover crop, but if nutrients are deficient, as can be measured by soil tests, it may pay to fertilize the cover crop with part of the fertilizer destined for the next primary crop.

Nitrogen is the element that most often is too low to provide maximum growth of nonleguminous cover crops. Leguminous cover crops also may have insufficient N if they were not previously grown on the particular field and if the seeds have not been inoculated with a suitable strain of nitrogen-fixing bacteria.

Rotations

Crop rotation often can be used to increase OM and SOM. The increase in organic matter is accompanied by several advantages. Growing more than one primary crop increases the diversity of soil organisms, which favors control of harmful insects, fungi, bacteria, nematodes, and weeds. The inclusion of legumes in the rotation helps supply needed N, allowing considerable savings in fertilizer costs. Using crops in the rotation that require limited cultivation lessens erosion, which in turn favors the accumulation of SOM.

A number of rotations are listed in Chapter 6. Choosing a suitable rotation needs to take the following items into consideration: (1) the suitability of the rotation for the soil and climate; (2) whether it provides for special needs such as nematode, disease, or weed control; (3) whether available machinery and labor can handle the new crops efficiently; and (4) whether the new primary crops can be marketed profitably.

Although rotation with primary crops may not be suitable for many farms, very few farms cannot benefit from rotations with cover crops grown in off periods (late fall-winter-early spring or between crops).

Crop Residues

As indicated in Chapter 6, residues can be important contributors of OM. An effort should be made to increase amounts produced and to utilize them to full advantage. Residues increase as growing conditions (nutrients, pest control, moisture, and soil aeration) for the primary crop are optimized. Removal or burning of residues should be avoided if at all possible. Residues are most effective if left on the surface to provide erosion protection during the winter months and more favorable temperatures and moisture conditions during the growing period. Planting through the residues is possible with special planters equipped with coulters. For poorly drained soils, it may be necessary to work the residues into the soil, but this should be done with implements that leave a major portion of the residues on or near the surface. More of the residues can be left on the surface by (1) preparing a part of the soil early so that it is suitable for planting in the spring, but leaving the remainder to provide good erosion control (zone tillage); or (2) using more or less permanent ridges that are high enough to provide a suitable bed for planting (ridge tillage).

Using OM Not Produced in Place

Many other sources of OM are available, but since they are not grown in place they may not be economical for many commercial growers. Nearly all of these are suitable for landscaping purposes and home garden use. Some of them are economical for commercial growers of high-priced items, such as flowers, foliage, ornamental trees, herbs, and some vegetables and fruits. Economic suitability also varies with the cost of the OM and the cost of putting it on the field. Much of the latter depends upon the distance the material has to be hauled and whether its handling can be mechanized. A number of materials were covered in some detail in Chapter 7. Some of the more important items are reviewed here.

Manure

Produced on the farm. A great deal of manure is produced on the farm and needs disposal, but in disposing of it, growers can supply valuable nutrients that they do not need to buy as fertilizers. The amount of nutrients contributed by manure varies with the animal source, feed rations, and bedding and is closely related to its handling, some of which is outlined in Chapter 7. Proper handling enhances its nutrient content, increasing its economic value and making its use economically justifiable, especially if collection and spreading can be largely mechanized.

Greater efficiency can be obtained by soil and manure testing. Soil testing allows the grower to place the manure on fields that can better utilize the nutrients and possibly avoid potential pollution problems. Testing manure provides a means for evaluating the amount of nutrients supplied. But to obtain reliable figures as to amount of nutrients applied, spreading equipment must be calibrated.

Better use of manure can be obtained by modifying the common practice of scrape and spread (daily removal and spreading of manure) and applying the manure to the closest fields. The present practice of spreading it on fields close to the barn tends to apply too much to certain fields and not enough to others. The net result is considerable waste of nutrients with potentials for polluting the overmanured fields and not applying enough nutrients to outlying fields.

Better distribution can be obtained by a combination of storing the manure and providing better ways of moving it to the fields. Liquefying the manure makes mechanical handling and spreading a simpler process. Two systems of liquid manure storage that lend themselves to more orderly distribution are available. (1) Housing the animals in buildings with slotted floors, through which manure can be flushed into a pit beneath the floor, allows the manure to be stored for three to six months. (2) Manure can be stored longer in aboveground storage tanks, concrete manure pits, earthen storage units, or mechanically aerated lagoons. The long periods of both indoor and outdoor storage permit more orderly transfer to fields. Outside storage allows much longer storage periods, but aboveground storage tanks are expensive and difficult to maintain. Concrete manure pits and earthen storage units are cheaper but may have more odor problems. Mechanically aerated lagoons require floating surface aerators that solve the odor problem but are expensive to install and operate.

The liquid manure can be more efficiently moved to the fields by using nurse tanks, large-volume hauling tanks that also can be used as spreaders, or portable aluminum piping that feeds the manure to big-gun traveling sprinklers that spread it. There are problems with equipment costs of these different approaches, but all offer reduction in application costs as well as providing better distribution of the manure (DePolo 1990).

Not produced on the farm. Manure not produced on the farm is another matter. It still may be economical for many growers if the source is nearby and if it is used for growing higher-priced items, especially if the manure is supplied at little or no cost. Growers faced with these options need to calculate the cost of nutrients, including cost of transportation and spreading low-analysis materials as compared to high-analysis fertilizers. But in addition to nutrients, manure supplies other advantages, the value of which are difficult to quantify. The OM, aside from its nutrient content, will have different values for different crops and different soils. The higher value for high-priced crops has already been alluded to, but a much higher value can also be placed on manure if it is to be used on a soil low in SOM that has problems, compared to a soil with ample SOM and no problems.

Unfortunately, there appears to be a lack of scientific data on which to base these other values. For many growers, the use of manure produced off the farm will be worthwhile if the value of the nutrients contained is no more than twice that purchased as fertilizer. In calculating fertilizer costs, it is desirable to compare at least half of the N supplied in manure with the cost of slow-release N, such as sulfur-coated urea (SCU).

It is unfortunate that much of the manure produced in animal factories is too far from commercial farming operations to be used extensively. This manure must be moved lest it cause serious problems by contaminating aquifers of potable water. One way that it can be handled is to compost it. Even though composting helps concentrate and dry manure, compost produced great distances from commercial farming operations probably will not be economical to use on these farms, but could be a good source of nutrients for homeowners, landscape gardeners, and various other urban needs, such as parks, roadsides, playing fields, and cemeteries.

Another potential problem with the use of manure is the possibility that it may carry organisms, such as *E. coli,* that can seriously affect human health. The organism can cause problems as it enters the food chain directly, or indirectly by contaminating water supplies. Although farmers have been using manure for centuries to grow food crops, apparently without causing serious problems, the current strain of *E. coli* may be much more virulent and could seriously affect human health. It may be some time before the problem can be evaluated. It could seriously limit the use of manure to nonfood crops or compost if it is to be used for food production. In either case, it would curtail OM additions for commercial agriculture, making it more dependent on residues and cover crops to sustain SOM.

Composts

Compost, while used for many years by homeowners and farmers of small acreage such as organic growers in the United States and a number of farmers in developing countries, has not been used extensively by large-scale growers, primarily because of the relatively high costs of its preparation and handling. Recent prohibitions on dumping yard waste in landfills has forced many municipalities to turn to composting to get rid of these wastes. The result is that a great deal of compost is now available at little or no cost, making it practical—for at least some commercial growers of high-priced items. As with manure, its economic value depends on the distance the compost has to be moved, its nutrient content, the value of the crop for which it will be used, and the value of qualities other than its nutrient content. Compost can have extra value as a mulch by suppressing disease, replacing herbicides with its smothering effect, and greatly enhancing temperature and moisture conditions for the primary crop.

Recent findings indicate that composts prepared properly have considerable disease suppression value. Proper composting largely gets rid of weed seeds and toxic substances, such as pesticides and allelochemicals. Machinery is now available that dispenses with much of the labor formerly used to prepare composts. All of these factors make the preparation of composts on large commercial farms more attractive, especially if the farm has some manure as a source of N to combine with various wastes with wide C:N ratios, such as tree trimmings, spoiled hay, or corn cobs.

Biosolids

Biosolids or sludge, with contents of nutrients suitable for plants and the need for large-scale disposal, make it a potential source of OM, especially for farms close to cities that need to dispose of these wastes. They no longer can be dumped in most bodies of water, and disposal in landfills is being phased out. Despite the huge potential, use of biosolids on a large scale for agriculture varies tremendously among states. According to figures for the year 2000, several states

(Arizona, Colorado, Indiana, Iowa, Oregon, and Wyoming) disposed of 90 percent or more of their combined annual total of 239,000 tons of dry solids by land application, but Connecticut, Hawaii, Massachusetts, New Jersey, New York, and Rhode Island, with an annual production of 990,000 tons of dry biosolids, disposed of less than 10 percent of their production on land (Goldstein 2000).

Several factors may account for relatively poor use of biosolids for agriculture in some areas. The potential dangers of disease, flies, and high metal contents have prompted regulations limiting their use. According to a recent survey, 16 states have one or more counties and/or towns banning or limiting use of biosolids for land applications, but, despite these limitations, land use has been holding rather steady. This may change, as the EPA has commissioned the National Academy of Science to evaluate the risks associated with using biosolids (Goldstein 2000). In the meantime, the use of biosolids has been set back by the refusal of certain food store chains to purchase vegetables produced with biosolids.

Until there is a definitive evaluation of the risks, there may be relatively few changes in the amounts of biosolids used for agriculture. For growers close to a good source, it may be worthwhile to use biosolids as an OM source. Assuming that it is economical, there should be no limitations from a health standpoint at present on their use for nonfood items, such as ornamentals, fibers, or possibly forages. Their use for producing food should be evaluated from the standpoint of heavy metal content and whether any limitations are imposed on its use, either by governmental agencies or purchasers of the product.

Peat

The high cost of peat eliminates it as a major source of organic matter for most farming operations, but it is highly useful for landscape purposes and for golf courses, mostly for its physical attributes. Considerable quantities are also used to make potting mixtures suitable for growing a wide variety of plants, although it is being increasingly replaced by composted wood because of costs.

Food-Processing Wastes

The large quantities of food processing wastes and their usual close proximity to commercial farms could make these wastes an important source of OM. The materials vary tremendously in pH, salts, and nutrient content, making it important to analyze sources before using them. Also, some materials may have a high COD (chemical oxygen demand) because of moisture content, making them difficult to handle and undesirable to add to soils unless some time elapses between application and planting, lest the material unduly rob O_2 from the developing primary plant.

Wood Products

Large quantities of wood wastes are available for increasing OM applications, but the wide C:N ratio of most of these materials precludes large usage for most commercial farm operations. There is a ready market for them in landscaping and home use, as these materials make excellent mulches around trees and shrubs and can be used to open up clay soils before preparing them for various landscape uses.

Composted wood wastes can be used for purposes other than potting mixes. Theoretically, there is a place for the compost as mulches for high-priced horticultural crops, where properly prepared compost could be useful in suppressing several plant diseases and could also reduce the need for herbicides. Part of the problem in making wood compost economical is the need to introduce enough N to provide narrower C:N ratios. This can be done with added fertilizer N, but it would make much more economic sense if the N additions could come from some other waste product, such as biosolids or manure.

Maximizing OM Additions

Increasing the amount of added OM enhances the benefits of SOM and helps ensure its buildup. Two methods of maximizing OM additions are (1) growing the cover crop as long as possible, and (2) keeping or placing the OM as mulch.

As pointed out in Chapter 6, it is desirable to grow the cover crop as long as possible, but doing so may cause moisture and nutrient shortages for the next primary crop. It is important to avoid such problems by irrigation and addition of missing elements. If recourse methods are not available, the cover crop will have to be terminated before water or nutrients are reduced to the point that they can hinder the next primary crop.

In Chapter 8, it was stressed that OM needs to be kept on the surface as mulch since it provides maximum moisture retention, temperature regulation, erosion control, OM preservation, and control of insects, disease, and weeds. But living mulch needs to be carefully controlled by selection, mowing, cultivation, and herbicides to avoid competing with the primary crop for light, moisture, and nutrients. In some cases, water and nutrients, strategically placed, can overcome some of the competitiveness of the living mulch. As indicated in Chapter 9, keeping residues and OM additions as mulch (NT) may have its drawbacks, particularly on poorly drained soils, unless some modification of NT such as zone tillage, stubble mulching, or ridge tillage is used. If these alternative methods cannot be used, OM may have to be incorporated, but it should be done with tools that leave most of the residues on or near the surface.

SUPPLYING NEEDED NUTRIENTS

The nutrients that need to be applied to grow the primary crop can be quite different for the COM system than for CT systems. As pointed out in Chapter 10, the COM system tends to improve the availability of P, K, and possibly most micronutrients but may reduce N, at least during certain stages of the primary crop. The decreased N availability occurs at times despite the N carried by legumes. Some of the problem of N shortage appears to be due to greater losses of N from leaching and denitrification resulting from the greater amount of moisture retained by the mulch. But a great deal of N shortage at certain stages of the primary crop's growth appears to be due to wide C:N ratios in the OM and poor release of organic N during cool or overly wet conditions, which often coincides with the time when leaching and/or denitrification occur.

The problems of materials with a wide C:N ratio and cool temperatures usually occur soon after the primary crop is planted, but excessive moisture can occur at any time during the primary crop's growth. Wide C:N ratios can also be aggravated by lack of moisture. If irrigation to correct the problem is not available, the cover crop needs to be terminated before moisture becomes inadequate.

These problems often require larger N inputs than with CT. The need for extra N appears to be greater when first starting COM, evidently lessening as the system tends to increase SOM with time. Rather than increasing N for COM blindly, we believe considerable savings can be made and the chances of polluting ground or surface waters can be reduced if the decision to add extra N would be based on soil tests for available N made shortly after the crop emerges.

Soil and leaf analyses are needed for COM as well as for conventional modes of farming to evaluate fertility programs. The idea that less testing is needed for COM because of the larger amounts of organic matter and the inclusion of legumes is faulty and can cost the grower. Actually, postplanting soil testing may be more important because of N problems.

If soil cannot be tested for available N (ammoniacal and nitrate nitrogen) shortly after planting, it is desirable to have some readily available N close to the roots of rapidly developing annuals so that plants can get a good start. A high-N transplanting solution is suitable for transplants. Banding 15-20 lb/acre of the scheduled N 2 inches from seed and 2 inches deeper than seedline at time of planting or applying it through drip irrigation about 10 days after seeding also works well. For low-cation-exchange soils, where lack of K also can be a problem due to leaching, the senior author has found that a 23-0-22 fertilizer made from equal amounts of ammonium nitrate and potassium nitrate and applied through drip irrigation has given good results.

CONTROLLING PESTS

As has been pointed out, pest control for COM can be different than that for CT. Not only can pests be different, but the amount of control needed for COM can vary tremendously depending on the primary and cover crops, amounts and kinds of added OM, and vagaries of climate and soil.

Need for IPM

Because pest problems vary so much, integrated pest management (IPM) is probably more important with COM than with conventional agriculture. Scouting is especially needed to determine whether pesticide applications are warranted for a particular situation because (1) there are tremendous differences in pest problems between various components of COM, and (2) the COM system is also much more dependent on beneficials for pest control than conventional agriculture. Scouting determines not only numbers and kinds of pests but whether beneficials are present and in sufficient numbers to control the pest. At times, pest outbreaks may be so severe that beneficials cannot provide economical control, and it may be necessary to use sprays or other measures on a temporary basis to control the pest. Frequent evaluations are necessary to determine whether treatment is needed. Once treatment has controlled the outbreak, scouting is needed to evaluate whether sufficient beneficials remain or need to be increased to adequately handle future problems.

An effective IPM program depends upon a monitoring system that follows pest trends to determine whether pesticide application is necessary. The effectiveness of the system largely depends upon the ability of the scout to

1. distinguish the pest and the beneficials, if available, at various stages of development;
2. maintain the frequency of examinations necessary to keep abreast of the situation;
3. be knowledgeable as to the stages of the pest that are damaging, and stages of the beneficials that are most sensitive to treatment;
4. know at which stage treatment is most effective in controlling the pest but least damaging to the beneficials;
5. know the various treatments that can accomplish this end; and
6. know all necessary inputs for growing the crop and how these inputs or management practices affect the pest and beneficials.

This is a large order, but usually most professional scouts are well trained and have basic knowledge that allows them to quickly evaluate new conditions so that they can fulfill their goal. It helps if the scout, in addition to being highly observant, inquisitive, and capable of keeping good records, also keeps abreast of the scientific literature and new developments in the industry so that he or she can quickly adopt new techniques that further the profession.

Professional scouts that can supply good IPM services are available in many parts of the United States. The cost of the service varies depending on several factors, such as the education and experience of the scout, efficiency or management of the company providing the service, crop and pests involved, and the frequency of visits needed to keep abreast of the situation. In a case study of IPM in Florida, it was found that costs of the services, used primarily on vegetable farms, were more than made up in savings on insecticides but not in use of bactericides, fungicides, or nematocides. Savings on pesticides mostly were in the range of $200 to $400 per acre, but in some cases, no overall savings occurred but costs of IPM were largely met by the reduction in pesticide use. In calculating costs, no attempt was made to include the cost of the extra insecticides upon applicators, farmhands, and damage to the environment, particularly aquifers. Damage to aquifers from pesticides has already become a major concern in Florida, with its porous soils and high rainfall (Fisher 1990).

Scouting in this study was done either by firms providing pest monitoring service or by in-house scouts that were trained for the purpose. The firms providing the service may have an edge because the examination of many farms in the area may provide some knowledge as to the development of the pest, alerting the grower of the need for providing extra management to better prepare for the potential buildup. But the effectiveness of the two approaches often will depend on how well each has been trained and how much time is allowed for the diagnostic approach.

If in-house scouts are used, the grower must be certain that the individual selected has the temperament and inquisitiveness to carry out the IPM program as well as sufficient background to diagnose pests and beneficials. The minimum tools needed are a pocket lens of 10 to 20X power, sweeps and ground cloths, nets to catch insects, soil profile tubes, spade and/or trowel, pocket knife, tweezers, scalpel, collection vials, paper bags, field identification guides, and a sturdy notebook for records. Use of sticky cards or traps, some of which can contain lures or pheromones, can be helpful in diagnosing insects and beneficials (Wolf 1996).

Insects

One of the advantages of COM is the ease with which beneficials can be maintained. The combination of increased residues on the surface, increased SOM, reduced pesticide applications, and the almost continuous presence of cover crops favors the preservation of beneficials, both parasites and predators, that help control unwanted insects. The increased availability of beneficials with COM also helps reduce the need for pesticides. Additional reductions in pesticide applications are possible partly because increased OM and SOM promote biological diversity, keeping many pests in check.

To reap the full benefits of reduced pesticide applications under COM, it is necessary to promote beneficials as much as possible. Important approaches in achieving this aim are

1. knowing which beneficials are needed to control the pest;
2. using IPM to determine whether needed beneficials are present in sufficient quantities to do the job;
3. using IPM to limit application of pesticides to minimum needed;
4. using cover crops that can be helpful in maintaining beneficials; and
5. providing suitable cover crops that favor the beneficials during the growing season of the primary crop.

The increased use of cover crops, not only between crops but also as windbreaks and at field edges during the primary crop's season, tends to provide a haven for beneficials during the entire year.

Insect control for the COM system can employ tactics used by conventional agriculture, such as resistant cultivars, GM plants, and insecticides. Insecticides should not be applied until IPM decides there is a need and identifies the pest. As has been pointed out, insecticides should be limited wherever possible to those that are specific for the pest and decompose quickly in order to limit as much as possible a buildup of resistance to the insecticide or damage to beneficials and environment.

Disease

Disease control for COM can use many of the methods that are effective for CT, such as resistant cultivars, IPM to designate a need for fungicides or bactericides, and weather forecasting services to predict the need for fungicides and bactericides.

Much disease control is still based on routine periodic application or so-called insurance sprays, rather than spraying on an as-needed basis. Changing from an insurance to an as-needed basis, which is now possible with a scouting program and weather forecasting, can save considerable pesticide applications, although at times it may call for additional sprays. For the weather data, such as rainfall, temperature, and dew period, to be meaningful, it needs to be correlated with infection periods, or spore release of the pathogen (Stephens 1990).

Computer forecasting systems based on the relationship between infection period or spore release and environmental data have been devised and put to good use. Some typical forecasting services in use are Tom-Cast (for tomato diseases in Ohio), MARYBLYT (for apple blight), MELCAST (for melon diseases), and Ventem (for apple scab). A number of compact weather stations are now being offered that are useful for on-farm evaluation of weather conditions that might predispose plants to bacterial or fungal infections. These stations, suitable for both CT and COM, are capable of logging data for several conditions, such as leaf wetness, relative humidity, temperature, and rainfall, that are closely related to disease development. A grower using weather data collected at the farm and combining it with a forecasting service can greatly improve the effectiveness of a spray program.

Disease control for COM, while having similarities with that for CT, may have advantages allowing less fungicide or bactericide to be used. As pointed out earlier, the biological diversity fostered by COM tends to keep many disease organisms in check. The improved soil porosity tends to reduce levels of several soilborne diseases that seem to get the upper hand as plant roots fail to get enough O_2. Better plant vigor also tends to provide better disease resistance. An important advantage of COM over CT for controlling disease relates to reducing the splash effect and reducing tillage. By reducing splashing, fewer microorganisms are spread on plants. Reducing tillage reduces both splashing and the direct spreading of disease-causing microorganisms by equipment. The reduced tillage also results in less mechanical injury to plants, thereby providing fewer entry points for microorganisms.

Compost mulches, in addition to reducing splashing, can help suppress disease, but the compost must be fully developed (see Compost in Chapter 7). Anyone using composts for disease suppression should insist on acquiring a fully aged material.

Biocontrols based on agents that are a part of the biological diversity fostering disease control are already being put to use in disease control programs and offer considerable promise for the future. *Pseudonomas fluorescens* to control damping off of cotton caused by *Pythi-*

um and *Rhizoctonia* spp., and *Trichoderma* spp. that control several soilborne diseases have been used for several years (Stephens 1990).

Nematodes

It is generally recognized that if soil nematode populations are not excessive when starting COM, chances are that populations will not become a problem. Evidently, nematodes are kept in check by rotations and by biological diversity due to the increased organic matter. The higher moisture levels in soils with COM could also be contributing.

Weeds

It is possible to control weeds in the COM system with the same herbicides that are commonly used for CT, but control methods by mechanical means that are commonly used with CT generally are not applicable. Fortunately, several other control methods are available for COM that make practical weed control possible.

Mulches offer effective control of many weed problems, but need to be heavy enough to suppress weed growth. Mulches derived solely from primary crop residues may not be sufficiently deep to smother weeds, but their effects can be increased by proper spreading of the residues and by also using a preemergence herbicide. To obtain satisfactory weed control from mulches derived from cover crops, it may be necessary to increase the cover crop yield by adding nutrients, increasing seeding rates, using irrigation (if necessary), and delaying the termination of the cover crop. Mulches from composts can supplement sparse coverage from residues or cover crops, but may be economically feasible only for certain high-priced crops. In calculating costs of adding composts or any other aid, such as nutrients or irrigation, it is necessary to include its value for producing the primary crop. For cured compost, its disease suppression value, elimination of weed seeds, and reduction in pesticides and allelochemicals need to be calculated in the formula.

Reducing herbicides has a double advantage for both CT and COM by reducing costs and lessening chances for pollution. Normally, growers need to abide by prescribed rates, but the EPA does allow reduced rates if they are based on research data. Experiments show that it is possible to use lower rates of oxyfluorfen and bromoxynil for early-season control of several weeds in muckland onions, and lower amounts of glyphosphate for quackgrass control. Lower amounts of glyphosphate can also be used to terminate rye cover crops to be used as mulch for NT systems.

The effective rate appears to be related, at least for glyphosphate, to the volume of water used. As shown in Table 11.1, control of quackgrass was as effective with 0.55 kg·ha^{-1} as with 1.1 kg·ha^{-1}, providing volume applied was reduced from 300 L·ha^{-1} to 83 kg·ha^{-1}. The complete control of rye was also possible with about half as much glyphosphate if the volume applied was cut almost in half (Putnam 1990).

As with controlling insects and disease, IPM can reduce herbicide use both for CT and COM. IPM can result in considerable herbicide savings by applying the herbicide when weeds are young and relatively easy to kill. Another saving can occur by limiting applications to spot treatments before the weed has spread.

TABLE 11.1. The effect of glyphosphate rates and volumes on quackgrass control and rye kill

Glyphosphate rate (kg·ha^{-1})	Control rating*		
	Liters applied per hectare		
Quackgrass	83	166	300
0.55	7.3	6.0	4.8
0.82	9.3	6.8	6.5
1.1	9.8	7.5	7.5
Rye	83	124	300
0.18	6.5	5.3	1.8
0.37	10.0	10.0	6.0
0.75	10.0	10.0	10.0

Source: Chase, W. R., A. R. Putnam, B. H. Zadstra, E. Hanson, J. Hull Jr., C. Peterson, and T. Wallace. 1986. Weed Control Field Research in Horticultural Crops. *Michigan State University Horticultural Report* 40:107. Cited by A. R. Putnam. 1990. Vegetable weed control with minimal herbicide inputs, *HortScience* 25: 155-158.
*0 = no control or kill; 10 = complete control or kill

ADDING WATER

The COM system offers more effective utilization of water by increasing the amount of water moving into the soil and stored in the soil, improving soil wetting, reducing losses by evaporation, and increasing uptake because of a large root system near the surface that is stimulated by the improved water level.

The better use of water with COM can be an important link to agricultural sustainability because insufficient water may be one of the first shortages affecting agricultural production. Problems already exist, especially in arid regions, because irrigation has increased while populations also have increased, placing extra demands on limited supplies. Thus less water is available for maintaining fish populations in rivers used for irrigation, and major aquifers supplying irrigation waters have been depleted to dangerously low levels. Also, the storage capacity of existing reservoirs has decreased due to silting in by eroded sediments.

Excessive soil salts introduced by irrigation may be the first manifestation of insufficient water. Salts of various kinds, but mainly bicarbonates (HCO^{3-}), carbonates (CO_3^{2-}), chloride (Cl^-), sodium (Na^+), calcium (Ca^{2+}), and magnesium (Mg^{2+}), are introduced with irrigation waters. These and many other salts derived from fertilizers, various amendments including those from organic sources, and numerous minerals found naturally in soils can accumulate in concentrations high enough to impede water uptake by plants.

Excess salts occur partly because plants remove large amounts of water but very small amounts of salts. Soils become increasingly salty if the excess salts are not removed by rainfall or irrigation. Problems can exist wherever irrigation is practiced but are more common in arid regions, especially if drainage is impeded.

The amount of salts tolerated by different plants will vary. A number of highly useful plants such as field beans, alsike and red clovers, apples, apricots, grapefruit, oranges, cabbage, lettuce, and potatoes are very sensitive to salts. Growth tends to be reduced if conductivity of a soil solution, a useful measure of total soluble salts, approaches 2 dS/m (dS/m = millimole per centimeter expressed as mmhos/cm). On the other hand, there is a large group of plants (grain barley, cotton, canola, bermudagrass, rhodes grass, coconut and date palms, asparagus, and garden beets) that are quite tolerant to salts, with growth not restricted until conductivity is above 6 dS/m. Several major crops (corn, flax, oats, rye, sorghum, soybeans, wheat, sweet clover, and tomatoes) are medium tolerant, with yields affected as conductivity exceeds about 4 dS/m. Slightly higher salt values than indicated can be tolerated if the soil has good organic matter contents since more water can be held by the soil (Wolf 1999).

Growing crops more tolerant to salts is one approach to utilizing soils with salt buildup, but unless enough water is added to leach out excess added salts, there will be a continuing need for changing to more tolerant crops. In time, there will be no suitable crops to grow. The net result could be another region lost to agriculture. History is replete with failed irrigation agriculture. One of the more famous failures, in the Tigris and Euphrates delta (the Fertile Crescent), resulted in a lost civilization, but not before farmers switched from growing wheat, a moderately salt-tolerant crop, to grain barley, a salt-tolerant crop.

Reducing damage from salt buildup is possible if excess salts can be removed by drainage. If water tables are too high to permit efficient drainage, a drainage system becomes an integral part of any irrigation system if salts are to be controlled (Hillel 1987).

The removal of excess salts is also dependent on adequate soil porosity, which allows ready movement of the excess salts. The COM system facilitates drainage of excess salts, but porosity may be adversely affected if irrigation water contains an excess of Na^+ as compared to Ca^{2+}, Mg^{2+}. The excess Na^+ favors soil dispersal, which reduces water infiltration and drainage by reducing porosity, whereas Ca^{2+} and Mg^{2+} tend to limit aggregate dispersal, preserving porosity. The sodium adsorption ratio (SAR), which can be calculated from the equation below, is a convenient means of evaluating irrigation water. Water that has an SAR less than 3 usually will not adversely affect porosity. Those sources with SAR values of 3 to 6 probably are safe to use, particularly if sufficient Ca^{2+} and Mg^{2+} are maintained in the soil cation exchange. Water with SAR values greater than 6 tends to create serious problems of drainage and should be avoided.

$$SAR = \frac{Na^+}{\sqrt{\frac{1}{2}\left(Ca^{2+} + Mg^{2+}\right)}}$$

where Na^+, Ca^{2+}, and Mg^{2+} are expressed as milliequivalents per liter.

The saturation of cations in the soil exchange complex has a bearing on the efficacy of irrigation water in maintaining satisfactory drainage. Soil cation exchange complexes saturated with about 70 percent Ca^{2+} and 15 percent Mg^{2+} and less than 5 percent Na^+ tend to provide good porosity with good drainage. As the percentage saturation of Ca^{2+} falls below 65 percent and the Na^+ saturation increases to 10 percent or more, porosity tends to decrease and there are increasing problems of infiltration and drainage.

Evidently, these percentage saturation values are useful, providing there is sufficient SOM to maintain satisfactory aggregate stability. The senior author has observed that gypsum applied to a large number of soils in Central America, the Caribbean, and coastal plain soils in the United States that maintained at least 70 percent saturation of Ca^{2+} were usually satisfactorily drained, but drainage and water movement in the heavier soils (loams, silt loams, and clay loams) was impaired as SOM fell close to 1 percent.

The COM system, by making better use of rainfall and helping to provide better drainage to remove excess salts, gives us an edge in achieving agricultural sustainability, but sustainability cannot be maintained unless we learn to use irrigation more efficiently. Sustainability can be increased by reducing the amount of irrigation water that is now being wasted. Much water is lost by poor storage, delivery, and application methods. Considerable irrigation water can be saved if applied only when crops need it, but this would require better scheduling and availability. Taking the politics out of water use and pricing irrigation correctly also could limit waste. Much of this is beyond the scope of this book, but the reader can gain some insight to the problem by referring to Hillel (1987).

EQUIPMENT FOR CONSERVATION TILLAGE

Terminating Cover Crops

Cover crops can be killed by herbicides, such as glyphosphate or paraquat, but in many cases, growers may prefer to do this mechanically. A number of cover crops can be terminated by such methods as mowing, rolling, flailing, chopping/flattening with a rolling stalk chopper, disking, undercutting, or plowing. Because it is desirable to have most of the organic matter on the surface, mowing or undercutting are preferred, but disking can be acceptable if not too much cover is worked into the soil. Plowing is not desirable and should only be used if there is a severe compaction problem or considerable stratification of pH and/or nutrients.

Some cover crops, such as annual ryegrass, can be killed by disking. The early bloom period is preferred for ryegrass. Other cover crops, such as barley, wheat, and field peas, can be killed either by disking or mowing. Oats, rye, and crimson clover can be killed by mowing. The preferable time for the oats is the milk or soft dough stage, that for wheat when it begins to flower, and crimson clover any time after bud stage. Some cover crops, such as cowpeas, are not killed by mowing or rolling but have to be undercut as well. Hairy vetch is killed by a number of mechanical methods, i.e., rotary mowing, flailing, cutting, subsoil shearing with an undercutter, or chopping/flattening with a rolling stalk chopper (Clark 1998).

Subsoiling

Although there is less compaction from equipment with conservation tillage than with CT, and increased SOM is an asset in slowing compaction, gravity and traffic tend to compact soil in time. The almost complete elimination of plowing and reduction of disking used with conservation tillage largely eliminates mechanical methods of periodically opening up soils, leaving most of the correction to subsoiling or an occasional plowing.

Subsoiling, or deep tillage with a chisel plow, can be used to reduce compaction. In most cases, it can be conveniently done at planting time by mounting a subsoiler ahead of the seeder but behind coulters that cut through the surface residue, or prior to planting using a subsoil bedder for ridge tillage (Figure 11.1). At planting time, it usually is desirable to move some of the loose soil that has been thrown out of the cut back into the opening. Compressing this soil slightly before seeding the crop and then compressing some more after seeding to form a firm seedbed helps ensure a good stand.

FIGURE 11.1. Subsoiler equipped with coulters opens compact soils while leaving much of the residue on the surface. (Photo courtesy of Adrian Puig, Everglades Farm Equipment, Belle Glade, Florida, and Matthew A. Weinheimer, John Deere Des Moines Works.)

Subsoiling of many soils is required annually, since the effects seldom last more than about eight months (Wright et al. 1980). The best time to subsoil is when the soil is fairly dry and it tends to shatter as the chisel is drawn through it. Unfortunately, soils at planting time in many areas may contain too much moisture for effective subsoiling. In such cases, and where soil is severely compacted, it may be necessary to delay the subsoiling until late summer after the primary crop is terminated.

The depth of the subsoiling depends upon the nature of the compaction. Because conservation tillage tends to provide a satisfactory moisture horizon near the surface, opening up about a foot of soil with the chisel plow often will be sufficient. For some soils that have compaction problems at deeper layers, it is worthwhile to go deeper (18-24 inches). Going deeper should be avoided if not necessary since it requires considerably more power, wider tires, and heavier tractors, all of which increases compaction under the wheels and in adjacent areas. It has been suggested that a heavy tractor tire should not pass any closer than 25 cm (10 inches) from planting rows or effectiveness of the subsoiling can be reduced, and a tire wider than 50 cm (20 inches) can limit the beneficial effects of subsoiling in a 100 cm (40 inches) row spacing (Throckmorton 1986).

Planting Through Residues

Some of the problems of planting in conservation tillage are outlined in Chapter 10. Most of the problems involve placing the seed at the proper depth, covering it with the proper amount of soil, and then compressing the soil to get proper contact of seed and soil. Proper placement and contact of seed and soil are necessary for sufficient moisture to be available for the germinating seed and to reduce damage from birds or rodents.

The planting process is complicated by the presence of large amounts of residues, but they usually can be handled by using coul-

FIGURE 11.2. A smooth coulter (knife) and opening mechanism designed to cut through heavy residues and open compact soils. The system, consisting of a DMI (DMI, Inc., Goodfield, Illinois) and an ACRA-Plant (ACRA Plant Sales, Inc., Garden City, Kansas), was designed for a transplanter (Figure 10.7) but should work well placed before seeders. (Photo courtesy of Dr. R. D. Morse, Virginia Polytechnic Institute and State University, Blacksburg, Virginia.)

ters or offset double disk openers to cut through the trash (Figure 11.2). Trash cleaners to clean a small strip (6-10 inches) ahead of the planter may be necessary for some planting operations where very heavy concentrations of residues are present. In some cases, such as heavy corn stalk residues, it may be necessary to use stalk choppers ahead of the planting process to effectively solve the problem.

Stalk choppers may be needed for planting with combination tillage/seeders. Although combination tillage/seeders are designed to plant through trash by providing extra clearance for residues through the use of fore and aft placement of chisels, heavy residues can clog the operation.

Tillage/seeders are capable of performing shallow tillage and seeding in one operation. The preparation of the soil is generally similar to that accomplished by field cultivators. In some cases, shanks are appreciably longer and similar to chisel plows, offering some subsoiling with the planting process. Seeding is air activated, blowing the seed from metering heads and allowing it to drop behind the shanks mounted on rigid bars mounted to the frame. Row spacings are varied by blocking some seed delivery. Although there are compensations for seeding in uneven soil, seed depth usually is not as accurate as with drills using independently mounted seed openers. Generally, tillage/seeders offer some savings in tillage operation and have more flexibility in application of fertilizers and pesticides, but tend to have problems with heavy plant residues (unless stalk choppers are used) and obtaining accurate seed depth and coverage.

Proper adjustment of both regular planters and tillage/seeders at each field is a necessity in order to plant through variable kinds and amounts of trash. Failure to do so can compromise stands and early growth.

Application of Pesticides

As pointed out previously, COM in combination with IPM can result in considerable savings on pesticides. Nevertheless, sprayers will be needed from time to time to apply pesticides for insect, disease, and weed control. Sprayers used for weed control also can be used to apply herbicides to terminate cover crops or sods.

Accurately calibrating sprayers prior to their use is even more essential for COM than for CT, partly because COM emphasizes using

FIGURE 11.3. Hooded sprayer. Spraying inside a hood over primary crop row (left) permits spraying it with fungicides or insecticides with little or no injury to beneficials on cover crops in aisles. By centering the spray nozzle between the hoods (right), herbicide application can be directed to the cover crops in aisles to suppress or terminate it without damage to the primary crop which is protected by the hoods. (Photos courtesy of Steve Clausen, Redball, LLC, Benson, Minnesota.)

the smallest amount of pesticide that is effective, not only to reduce costs but to lessen impact on beneficials.

For the same reasons, growers must use techniques that maximize the accuracy of pesticide deposition. Frequent calibration combined with frequent adjustment and replacement of faulty nozzles can be helpful, but methods such as electrostatic or controlled-droplet application may be necessary to maximize economy of pesticide application.

Reduction of pesticide drift by use of proper nozzles and their placement, correct pressures, and proper hooding can also aid in reducing the amount of spray used, but even more important are essential for the maintenance of beneficials (Figure 11.3). Proper hooding is also essential to prevent damage to the primary crop when applying herbicides to control weeds or cover crops. Frequent examination of hoods, especially just prior to starting an application of herbicide, can pay big dividends in reduced damage to the primary crop.

Summary

There has been a growing concern for some time that intensive agriculture is not sustainable because of its use of large amounts of synthetic fertilizers and pesticides and increased dependence on irrigation.

Some of the early criticisms of practices that increase erosion, dangerous pesticide applications, and use of artificial fertilizers that fail to supply adequate nutrition have been blunted by changes, many of which have been mandated by law. Practices promoted by the Soil Conservation Service, established in 1935, have reduced some of the erosion problems. Changes induced by pesticide regulation, starting with the passage of FIFRA in 1947 and modified in 1954 and again in 1972 with the establishment of the EPA, have greatly relieved some of the pesticide problems. The inadequacies of early synthetic fertilizers have largely been overcome by adding micronutrients to them and coating some of the N fertilizers so they do not leach readily.

Beneficial as these improvements have been, they have not completely eliminated criticism of the modern intensive mode. The criticisms focus on at least four items:

1. The intensive system, with its great dependence on added fertilizers, pesticides, and irrigation, accelerates the depletion of necessary elements of agriculture, reducing its sustainability.
2. Excessive use of persistent synthetic fertilizers and pesticides unduly pollutes surface waters and aquifers for long periods.
3. There is still too much erosion, although some of it may be due to the failure to fully comply with conservation practices, some of which are either too expensive or too difficult for many growers to maintain.
4. The practice of monoculture, with its low input of OM along with excessive tillage and lack of rotations, tends to deplete SOM, increasing erosion and pest problems.

Some critics have turned to organic farming as a means of making agriculture sustainable. Definitions of organic farming are variable and the term can have different meanings for different farmers. Essentially, it requires the almost complete abandonment of synthetic fertilizers and pesticides, depending on crop rotations, legumes and other cover crops, manure, compost, rock phosphate, minerals, rock dusts, and natural pesticides, such as oils, rotenone, or derris, to supply needed nutrients and to control pests. The organic farming system does provide certain key benefits, namely reduced erosion, increases in SOM, and considerable reduction of inputs.

The increased OM and better SOM may well be the keystone in the success of organic farming. The close association between organic matter and crop production evidently is due to the profound physical, chemical, and biological advantages that organic matter gives to most soils.

Unfortunately, organic farming as it is now practiced cannot provide sufficient food and fiber for future generations even if it does a better job of maintaining SOM. The deficiencies appear to be related to inadequate nutrients at certain stages of crop growth, inadequate control of pests, and insufficient reduction of tillage. In fact, the combination of intensive agriculture and organic farming are not adequately taking care of current needs, since it has been estimated that at least one-third of the world's population is not getting enough to eat. Although much of the current shortage may be due to political factors, turning completely to organic farming would appear to increase the shortage rather than decrease it, because organic farming appears to produce about 5 to 10 percent less food than intensive systems. Actually, some of the worst shortages are taking place where the majority of the farming is organic, and production usually improves (at least temporarily) as intensive systems are added. In all probability, current shortages will increase as world populations increase, since it appears that intensive systems cannot be sustained and organic farming cannot supply enough food and fiber.

The authors believe that much of these shortages can be mitigated and agriculture sustained over the long run by combining the best of organic farming with that of intensive agriculture. The basics of organic farming, which emphasizes rotations and adding OM, can help improve SOM. The use of conservation tillage over the past 20 to 30 years proves that it is possible to maintain and in many cases increase SOM.

The use of some intensive agriculture inputs can improve yields. Quick-acting synthetic fertilizers could well increase yields by supplying N and sometimes K and other needed elements when OM and SOM fail to provide these elements soon enough to take care of fast-growing plants. Yields, at times, can also be improved by the use of synthetic pesticides, which are not not permitted by organic farming. The "natural" insecticides approved for organic farming use may be too slow or inadequate to control severe outbreaks of pests, and the timely use of a synthetic pesticide can make a substantial improvement in salable crop yields.

The combination of conservation tillage, which provides for improved organic matter, with large additions of OM left on the surface (COM) offers the possibility of combining the best of organic and intensive farming and promises to be sustainable for long periods, providing suitable cultural methods are adopted to make the new system practical.

The essential cultural methods to make COM practical are as follows:

1. *Eliminate as much tillage as possible.* Avoid moldboard or disk plowing. Using coulters and chisel plows before seeders will provide sufficient aeration and reduce compaction sufficiently on most well-drained soils. On poorly drained soils it may be necessary to do additional tillage, either by forming ridges, adding some zone tillage, or using subsoilers. If full-width tillage is needed, use implements that leave appreciable residues on the surface.

2. *Add as much organic matter as possible.* Growing it in place in the form of sods, hays, forages, cover crops, and plant residues is the cheapest method of maximizing OM additions for most commercial farms. Utilizing primary crop residues and growing cover crops between primary crops are generally useful for most farms. Adding OM not grown in place, while not being economical for many commercial operations, can be helpful for some growers of high-priced items and for home and commercial landscaping. Manure, compost, biosolids, and several waste products may be suitable OM sources for some farms because of proximity to the sources and need for disposal. The availability of composts has greatly increased in recent years, making it more attractive for many farms.

3. *Rotate crops.* Avoiding monoculture provides for SOM increase and reduces pest problems. Include legumes as a primary and/or cover crop to provide free N in the rotation.

4. *Allow added OM to remain on the surface as mulch.* Residues, terminated cover crops, and composts used as mulch reduce upper soil temperatures, increase availability of water, and reduce erosion. Living mulches, primarily used for vineyards and orchards in the past but now also being used for row crops, also add OM and reduce erosion but may compete with the primary crop unless care is taken. Competition can be reduced by selecting cover crops that die back as the primary crop advances, or limiting the cover crop's growth by mowing, selected herbicides, or limited cultivation. Competition can be limited by replacing some water and/or nutrients used by the cover crop.

5. *Increase biomass of residues or cover crops by supplying adequate nutrients and water.* Increase biomass by extending the growing season (increasing seeding rate, starting the cover crop in the primary crop, and delaying termination until cover crop blooms). Early termination can be avoided by irrigation.

6. *Use appropriate machinery for conservation tillage.* Conservation tillage usually requires substantial changes in planter equipment. Existing equipment may be modified, although it may be desirable to purchase equipment especially designed for conservation tillage. Although conventional equipment can be used, most conservation farmers will find it useful to have hooded sprayers for herbicide applications and flail mowers or rollers for terminating cover crops.

7. *Alter fertilizer program to conform with conservation tillage needs.* Unduly decreasing fertilizer applications because of nutrients added as OM or released from SOM can lead to reduced yields. Because of greater losses of N by leaching and denitrification and tieup of N due to wide C:N ratios or the presence of lignins, there often is a need for additional N, especially during cool wet periods and early stages of conservation tillage. Rather than blindly adding extra N, base applications on needs as shown by rapid soil tests for NH_4-N and NO_3-N.

8. *Eliminate pesticides as much as possible.* Conservation tillage offers an entirely different approach for pest control, and considerable savings in use of pesticides can be made by utilizing the new system to advantage. Make use of beneficials to limit insect damage,

growing cover crops to help maintain them and avoiding pesticides that can unduly suppress them. Use mulches and/or composts wherever possible for weed control. Make full use of IPM and forecasting services to limit pesticide application. Avoid broad-spectrum or long-lasting pesticides, using only those that are compatible with maintaining the beneficials.

9. *Adjust irrigation program to conservation tillage.* Conservation tillage increases the amount of water available for crops. Water can be saved by applying it only when needed as indicated by a suitable irrigation scheduling program. The extra water associated with conservation tillage lessens damage from salt buildup, but to avoid catastrophes associated with earlier irrigation programs, greater emphasis needs to be placed on limiting salt applications from irrigation water and fertilizers, and draining away excess salts. Organic matter helps in reducing salts from fertilizers and aids in drainage of salts by keeping soils open.

Appendix 1

Common and Botanical Names of Plants

Common Name	**Botanical Name**
Alfalfa or lucerne	*Medicago sativa* L.
Alligator weed	*Alternanthera philoxeroides*
Almond	*Prunus amygdalus* Batsch
Apple	*Malus* sp.
Apricot	*Prunus armeniaca* L.
Asparagus	*Asparagus officinalis* L.
Avocado	*Persea americana* P. Mill.
Azalea	*Rhododendron indicum*
Banana	*Musa* spp.
Barley	*Hordeum vulgare* L.
Beans	
broad, fava, field, or horse	*Vicia faba* L.
lima, snap, or wax	*Phaseolus vulgaris* L.
velvet	*Stizolobium deeringianum*
Beets	
mangel wurzel	*Beta vulgaris* L.
sugar	*B. saccharifera*
table	*B. vulgaris* L.
Bird's-foot trefoil	*Lotus corniculatus* L.
Blackberry	*Rubus fruiticosa* Auct.
Blueberry, highbush	*Vaccinium corymbosum* L.
Broccoli	*Brassica oleracea,* Botrytis Group
Buckwheat	*Fagopyrum esculentum* (Moench)
Cabbage	*Brassica oleracea* L., Capitata Group
Chinese	*B. rapa* L., Chinensis Group
Canada thistle	*Cirsium arvense* (L.) Scop.
Canola or rape	*Brassica napus* L.
Cantaloupe or muskmelon	*Cucumis melo* L., Reticulatus Group
Carrot	*Daucus carota* L.

Cauliflower	*Brassica oleracea*, Botrytis Group
Celery	*Apium graveolens* var. *dulce* (Mill.) Pers.

Cherry
  black   *Prunus serotina* Ehrh.
  sour   *P. cerasus* L.
  sweet   *P. avium* L.
Chickweed   *Stellaria media* (L.) Cyrillo
Citrus   *Citrus* spp.
Clover
  alsike   *Trifolium hybridum* L.
  alyce   *Alysicarpus* Desv.
  balansa   *T. blanasae*
  berseem or Egyptian   *T. alexandrinum* L.
  bur   *Medicago denticulata*
  crimson   *T. incarnatum* L.
  hubam   *Melilotus alba* var. *annua*
  ladino or white   *T. repens* L.
  paradana balansa   *T. balansae* 'paradana'
  Persian   *T. resupinatum* L.
  red or mamouth   *T. pratense* L.
  southern spotted burclover   *Medicago polymorpha*
  subterranean or subclover   *T. subterraneum* L.
  white   *T. repens* L.
  white sweet   *Melilotus alba* Medik.
  yellow sweet   *M. officinalis* Lam.
Coffee   *Coffea arabica* L.
Corn or maize
  common   *Zea mays* L.
  pop   *Z. mays* var. *everta*
  sweet   *Z. mays* var. *rugosa* Bonaf.
Cotton   *Gossypium hirsutum* L.
Cowpea   *Vigna unguiculata* (L.) Walp.
Cranberry   *Vaccinium macrocarpen selender* ex Ait.
Crotalaria   *Crotalaria juncea*
Cucumber   *Cucumis sativus* L.

Dandelion   *Taraxacum officinale*

Eggplant   *Solanum melongena* L.
Endive or escarole   *Cichorium endiva* L.

Appendix 1: Common and Botanical Names of Plants 321

Fescue
  chewings   *Festuca rubra* L. ssp. *commutata* Guad.
  meadow   *F. pratensis* Huds.
  red   *F. rubra* var. *genuine*
  tall   *F. arundinacea* (Schreb.) Wimm
Fig   *Ficus carica*
Fir, Douglas   *Pseudotsuga menziesii* (Mirbel) Franco
Flax   *Linum usitatissimum* L.
Foxtail
  meadow   *Alopercusrus pratensis* L.
  tall   *Setaria* spp.

Grape   *Vitis vinifera, V. labrusca,* and hybrids
  Muscadine   *Vitis rotundifolia* Michx.
Grapefruit   *Citrus ×paradisi*
Grass,
  bahia   *Papsalum notatum*
  barnyard   *Echinocloa crus-gali* (L.) Beauv.
  bent
    colonial   *Agrostis capillaris* L.
    Rhode Island   *A. tenuis* Sibth.
    seaside   *A. palustris*
    velvet   *A. canina*
  bermuda   *Cynodon tanadon dactylon* (L.) Pers.
  bermuda, coastal   *C. dactylon* (L.) Pers.
  blue
    annual   *Poa annua* L.
    Kentucky   *P. pratensis* L.
  brome
    awnless or smooth   *Bromus inermis* Leyss
    fringed   *B. ciliatus*
    great   *B. diandrus*
    prairie   *B. uniloides*
  bluejoint   *Calamagrostis canadensis* Beauv.
  canary   *Phalaris* L.
  centipede   *Eremochloa ophiuoides* (Munro) Hack.
  creeping bent   *Agrostis stolonifera* L.
  crested wheat   *Agropyron desertorum* Schultes
  dallis   *Papsalum* L.

goat, jointed	*Aegilops cylindrica* Host.
harding	*Phalaris* L.
horse nettle	*Solanum carolinense* L.
itch	*Rottboellia exalta* (L.) F.
johnson	*Sorghum halepense* (L.) Pers.
needlegrass	*Stipa comata* Trin. and Rupr.
orchard or cocksfoot	*Dactylis glomerata* L.
pangola	*Digitaria decumbens* Stent.
para grass	*Panicum barbinode*
prairie	*Bromus uniloides* (Wild) Kunth
quack	*Agropyron repens* (L.) Beauv.
rhodes	*Chloris gayana*
ryegrass	
annual or common	*Lolium multiflorum* Lam.
perennial	*L. perenne* L.
sorghum-sudan	*Sorghum bicolor sudanese* (L.) Moench
sudan	*S. bicolor sudanese* (P.) Stapf.
St. Augustine	*Stenotaphrum secundatum* (Walte.) Kuntze
switch	*Panicum virgatum* L.
wheat, crested	*Agropyron desertorum* (Fish ex Link) Schult.
zoysia	*Zoysia matrella* (L.) Merr.
Guar	*Cyamopsis tetragonoloba* (L.) Taub?
Hairy indigo	*Indigofera hirsutum* L.
Hazelnut or filbert	*Corylus avellana* L.
Hemlock, western	*Tsuga heterophylla* (Raf.) Sarg.
Hensbit	*Lamium amplexicaule* L.
Honeylocust, moraine	*Gleditsia triacanthos* var. *inermis*
Horseradish	*Armoracia rusticana* P. Gaertn., B Mey, and Scherb.
Klamath weed or St. John's wort	*Hypericum perforatum*
Lamb's-quarter	*Chenopodium album* L.
Lemon	*Citrus limon* L.
Lentil, black	*Lens culinaris* Medik.
Lespedeza	*Lespedeza striata* (Thumb. Ex Murr.) Hook. & Arn.
Lettuce	*Lactuca sativa* L.

Appendix 1: Common and Botanical Names of Plants

Macadamia	*Macadamia ternifolia* F. Muell
Mandarin or tangerine	*Citrus reticulata* Blanco
Marigold	*Tagetes* L.
Medic	*Medicago* spp.
Mediterranean sage	*Salvia apthiopsis*
Milkweed, common	*Asclepias syriaca*
Millets	
German (foxtail)	*Setaria italica* (L.) Beauv.
Japanese	*Echinochloa frumentacea* (Roxb.) L.
pearl	*Pennisetum glaucum* L.
wild proso	*Panicum millaceum* L.
Milo, dwarf yellow	*Sorghum vulgare*
Mint	*Menthe arvensis*
Mustard	*Brassica juncea*
Nutsedge	
yellow	*Cyperus esculentus* L.
purple	*C. rotundus* L.
Oak	*Quercus* spp.
Oat	*Avena sativa* L.
Onion	*Allium cepa* L., Cepa Group
Orange, navel and valencia	*Citrus sinensis*
Papaya	*Carica papaya* L.
Parsley	*Petroselinum hortense*
Parsnip	*Pastinaca sativa* L.
Pea	
English	*Pisum sativum* L.
field	*P. sativum* var. *arvense* (L.) Poir.
southern or black-eyed	*Vigna unguiculata* L.
Peach	*Prunus persica* L.
Peanut, common	*Arachis hypogaea* L.
Pear	*Pyrus communis* L.
Pecan	*Carya illinoensis* L.
Pepper	*Capsicum annuum* var. *annuum* L.
Pigweed	*Amaranthus* spp.
Pine	*Pinus* spp.
Plum	*Prunus domestica* L.
Poison ivy	*Rhus radicans* L.
Potato	
Irish	*Solanum tuberosum* L.
sweet	*Ipomea batatus* (L.) Lam.

Pumpkin — *Cucurbita pepo*
Puncture vine — *Tribulus terrestris* L.
Purslane — *Portulaca oleracea* L.

Radish, oilseed — *Raphanus sativus* L.
Ragweed — *Ambrosia artemisiifolia* L.
Rice — *Oryza sativa* L.
Russian thistle — *Salsola iberica* Sennen & Pau
Rye — *Secale cereale* L.

Safflower — *Carthamus tinctorius* L.
Sesame — *Sesamum indicum* L.
Sesbania — *Sesbania aculenta* L.
Shattercane — *Sorhum bicolor* (L.) Moench.
Sorghum — *S. vulgare* Pers.
Sorghum-sudangrass hybrids — *S. bicolor* × *S. bicolor sudanese*
Soybean — *Glycine max* (L.) Merr.
Spelt — *Triticum aestivum spelta*
Spinach — *Spinacia oleracea* L.
Squash — *Cucurbita pepo* var. *melopepo* L. Alpf
Strawberry — *Fragaria* × *ananassa* Duchesne
Stylo — *Stylosanthes humilis* Kunth.
Sugarcane — *Saccharum officinarum* L.
Sunflower — *Helianthus angustifolius* and *annuus* L.
Sunn hemp — *Crotalaria juncea*

Tea — *Camelia sinensis*
Timothy — *Phleum pratense* L.
Tobacco — *Nicotiana tabacum* L.
Tomato — *Lycopersicon esculentum* Mill.
Trefoil
  bird's-foot — *Lotus corniculatus* L.
  yellow — *Medicago lupulina*
Triticale — *Triticcosecale rimpani* Wittm.
Tropical kudzu — *Pueriaria phaseolioides*
Trumpet creeper — *Campsisradcans* (L.) Seem. Ex Bureau
Turnip — *Brassica rapa* L. var., Rapifera Group

Appendix 1: Common and Botanical Names of Plants

Velvet bean or mucuna	*Mucuna deeringiana* (Bert.) Merr.
Velvetleaf	*Abutilon theophrasti*
Vetch	
cahaba white	*Vicia sativa* × *V. cordata*
chickling	*Latyrus sativus* L.
common	*Vicia sativa* L.
crown	*Coronilla varia* L.
hairy	*Vicia villosa* Roth.
purple	*V. benghalensis* L.
woolypod or lana	*V. dasycarpa* Ten.
Walnut	*Juglans regia*
Watermelon	*Citrullus lanatus* (Thumb.) Matsum & Nakai
Wheat	*Triticum aestivum* L.

Appendix 2

Abbreviations and Symbols Used in This Book

Abbreviation/Symbol	Word/Unit
a.i.	active ingredient
Al	aluminum
Al^{3+}	aluminum ion
As	arsenic
B	boron
C	carbon
(°C)	degree centigrade
Ca	calcium
Ca^{2+}	calcium ion
cc	cubic centimeters
Cd	cadmium
CEC	cation exchange capacity
Cl	chlorine or chloride
Cl^-	chloride ion
C:N	carbon:nitrogen ratio
CO_3^{2-}	carbonate ion
COM	combination of conservative tillage with mulch
Cr	chromium
CT	conservation tillage
Cu	copper
Cu^{2+}	copper ion
EC	electrical conductivity
(°F)	degree Fahrenheit
F	fluorine
F^-	fluoride ion

327

Fe	iron
Fe^{3+}	iron ion
ft	foot or feet
ft^2	square foot or feet
g	gram
gal	gallon
H	hydrogen
H^+	hydrogen ion
ha	hectare
H_2CO_3	carbonic acid
HCO_3^-	bicarbonate ion
Hg	mercury
H_2O	water
H_2S	hydrogen sulfide
HSO_4^-	hydrogen sulfate ion
in or "	inch
K	potassium
K_2O	potassium oxide or potash
lb	pound
meq/100 g	milliequivalents per 100 grams
Mg	magnesium
Mg^{2+}	magnesium ion
ml	milliliter or cubic centimeter
Mn	manganese
Mn^{2+}	manganese ion
mmhos	millimhos
Mo	molybdenum
MoO_4^{2-}	molybdate ion
N	nitrogen
Na	sodium
Na^+	sodium ion
NH_3	ammonia
NH_4^+	ammonium ion
NH_4-N	ammonium nitrogen
Ni	nickel
N_2O	nitrous oxide

Appendix 2: Abbreviations and Symbols Used in This Book

NO_2	nitrogen dioxide
$NO_2\text{-}N$	nitrite nitrogen
NO_3^-	nitrate ion
$NO_3\text{-}N$	nitrate nitrogen
NT	no tillage
O	oxygen
OH^-	hydroxide ion
OM	organic matter
oz	ounce
P	phosphorus
Pb	lead
P_2O_5	phosphorus pentoxide (phosphate)
ppb	parts per billion
ppm	parts per million
pt	pint
qt	quart
RT	reduced tillage
S	sulfur
SAR	sodium absorption ratio
Se	selenium
Si	silicon
SO_2	sulfur dioxide
SO_4^{2-}	sulfate ion
SOM	soil organic matter
UAN	urea ammonium nitrate
Zn	zinc
Zn^{2+}	zinc ion

Appendix 3

Useful Conversion Factors and Data

Column A	Column B	To obtain B multiply A by	To obtain A, multiply B by
Length			
inches	millimeters	25.4	0.0393
	centimeters	2.54	0.3937
	meters	0.0254	39.37
feet	centimeters	30.48	0.0328
	meters	0.3048	3.28
yards	centimeters	91.44	0.01094
	meters	0.9144	1.0936
Volume			
cubic centimeters**	fluid oz	0.0338	29.574
	pints*	0.0211	473.176
	quarts*	0.00106	946.342
	gallons*	0.00026	3785
fluid ounces	cubic centimeters	29.574	0.0338
	cubic inches	1.805	0.5541
	liters	0.0296	33.78
	pints*	0.0625	16
	quarts*	0.0313	32
	gallons*	0.0078	128
quarts	fluid ounces	32	0.03125
	pints	2	0.5
	liters	0.9463	1.0567
	cubic centimeters	946.358	0.0011
	cubic inches	57.75	0.0173

SUSTAINABLE SOILS

	gallons*	0.25	4
gallons*	pints*	8	0.125
	quarts*	4	0.25
	fluid ounces	128	0.0078
	liters	3.785	0.2642
	cubic inches	231	0.00433
	cubic feet	0.1337	7.481
	cubic yards	0.00495	201.974
cubic feet	liters	28.316	0.0353
	quarts*	29.922	0.0334
	gallons*	7.481	0.1337
	cubic inches	1728	0.00058
	cubic yards	0.037	27
	cubic meter	0.028	35.315
cubic yards	cubic feet	27	0.037
	cubic meters	0.7646	1.3079
	quarts*	807.9	0.00124
	gallons*	202	0.00495
cubic meter	cubic feet	35.7	0.0283
	gallons*	264.1721	0.0038
Weight			
ounces	pounds	0.0625	16
	kilograms	0.0283	35.27
pounds	kilograms	0.04536	2.205
	ton (short)	2000	0.0005
	ton (long)	2240	0.00045
	ton (metric)	2205	0.00045
Elements			
nitrogen (N)	ammonia (NH_3)	1.2159	0.8224
	ammonium	1.2878	0.7765
	nitrate (NO_3)	4.4261	0.2266
phosphorus (P)	phosphorus pentoxide (P_2O_5)	2.291	0.436
potassium (K)	potash (K_2O)	1.205	0.830

Area

acres	hectares	2.4709	0.4047
	square meters	4046.8564	0.000247
hectares	square meters	10000.0	0.0001

Rate

Pounds per acre	kilograms per hectare	1.121	0.892

Note: Temperature conversions: Degrees Fahrenheit = (degrees centigrade × 1.8) + 32; degrees centigrade = (degrees Fahrenheit − 32) × 0.56.
*U.S. fluid
**1 cubic centimeter = 1 milliliter

Bibliography

Abdul-Baki, A.A., J.R. Teasdale, and R.F. Korcak. 1997. Nitrogen requirements of fresh-market tomatoes on hairy vetch and black polyethylene mulch. *HortScience* 32(2):217-221.
Ahn, J.K. and I.M. Chang. 2000. Allelopathic potential of rice hulls on germination and seedling growth of barnyard grass. *Agronomy Journal* 92:1162-1167.
Akemo, M.C., M.A. Bennett, and E.E. Regnier. 2000. Tomato growth in spring-sown cover crops. *HortScience* 35:843-848.
Alabouvette, C., P. Lemanceau, and C. Steinberg. 1996. Biological control of Fusarium wilts: Opportunities for developing a commercial product. In R. Hall (Ed.), *Principles and Practices of Managing Soilborne Plant Pathogens* (pp. 192-212). St. Paul, MN: The American Phytopathological Society.
All, J.N. and G.J. Musick. 1986. Management of vertebrate and invertebrate pests. In M.A. Sprague and G.B. Triplett (Eds.), *No-Tillage and Surface Tillage Agriculture* (pp. 347-388). New York: John Wiley and Sons.
Aulakh, M.S., T.S. Khera, J.W. Doran, Kuldip-Singh, and Bijay-Singh. 2000. Yields and nitrogen dynamics in a rice-wheat system using green manure and inorganic fertilizer. *Soil Science of America Journal* 64:1867-1876.
Ayanlaja, S.A., S.O. Owa, M.O. Adigun, B.A. Senjobi, and A.O. Olaleye. 2001. Leachate from earthworm castings break seed dormancy and preferentially promotes radicle growth in jute. *HortScience* 36(1):143-144.
Aylsworth, J.D. 1996. No-till transplanter improves yields, cuts costs. *American Vegetable Grower*, April:18, 20.
Babcock, J.M. and J.W. Bing. 2001. Genetically enhanced Cry1F corn: Broad spectrum lepidoptera resistance. *Down to Earth* 56:10-15.
Bahe, A.R. and C.H. Peacock. 1995. Bioavailable herbicide residues in turfgrass clippings used for mulch adversely affect plant growth. *HortScience* 30:1393-1395.
Baker, R. and T.C. Paulitz. 1996. Theoretical basis for microbial interactions leading to biological control of soilborne plant pathogens. In R. Hall (Ed.), *Principles and Practice of Managing Soilborne Plant Pathogens* (pp. 50-79). St. Paul, MN: The American Phytopathological Society.
Barker, A.V., T.A. O'Brien, and M. L. Stratton. 2000. Description of food processing products. In J.M. Bartels and W.A. Dick (Eds.), *Land Application of Agricultural, Industrial, and Municipal By-Products*. Soil Science Society of America Book Series #6. Madison, WI: Soil Science Society of America.

Beare, M.H. 1997. Fungal and bacterial pathways of organic matter decomposition and nitrogen mineralization in arable soils. In L. Brussaard and R. Ferrera-Cerrato (Eds.), *Soil Ecology in Sustainable Agricultural Systems* (pp. 39-70). Boca Raton, FL: CRC Press.

Beneficial Bug Guide. 2000. Willoughby, OH: Meister Publishing Company.

Blevins, R.L., G.W. Thomas, M.S. Smith, and W.W. Frye. 1983. Changes in soil properties after 10 years of no-tillage and conventionally tilled corn. *Soil Tillage Research* 3:135-146.

Boosalis, M.G., B.L. Doupnik, and J.E. Watkins. 1986. Effect of surface tillage on plant diseases. In M.A. Sprague and G.B. Triplett (Eds.), *No-Tillage and Surface Tillage Agriculture* (pp. 389-404). New York: John Wiley and Sons.

Bordelon, B.P. and S.C. Weller. 1995. Cover crop effects on weed control and growth of first-year grapevines. 95th Annual Meeting, American Society for Horticultural Science, July 30-August 3, Montreal, Quebec, Canada.

Bradley, J.F. 2000. Why no-till and conservation tillage cotton? "Con-Till Cotton," A Special Supplement to Farm Chemicals, August. Willoughby, OH: Meister Publishing Co.

Brady, N.C. and R.R. Weil. 1999. *The Nature and Properties of Soils,* Twelfth Edition. Upper Saddle River, NJ: Prentice-Hall, Inc.

Broschat, T.K. n.d. Manganese binding by municipal waste composts used as potting media. *Tropicline* 4:1-4. [Tropical Horticultural Newsletter of the Ft. Lauderdale Research and Education Center and the Department of Ornamental Horticulture, Ft. Lauderdale, FL.]

Bruehl, G.W. 1987. *Soilborne Plant Pathogens.* New York: Macmillan Publishing Co.

Brundland, G.H. 1987. *Our Common Future. Report of the World Commission on the Environment and Development.* Oxford: Oxford University Press.

Brussaard, L. 1997. Interrelationships between soil structure, soil organism and plants in sustainable agriculture. In L. Brussaard and R. Ferrera-Cerrato (Eds.), *Soil Ecology in Sustainable Agriculture Systems.* Boca Raton, FL: CRC Press.

Brussaard, L. and R. Ferrera-Cerrato (Eds.). 1997. *Soil Ecology in Sustainable Agricultural Systems.* Boca Raton, FL: CRC Publishers.

Brust, G.E. 1994. Natural enemies in straw-mulch reduce Colorado potato beetle populations and damage in potato. *Biological Control* 4:163-169.

Bruulsema, T. and G. Stewart. 2000. Zone tillage: A fall option for corn. *Farm Chemicals,* September:42, 47.

Campbell, C.L. and D.A. Neher. 1996. Challenges, opportunities, and obligations in root disease epidemiology and management. In R. Hall (Ed.), *Principles and Practices of Managing Soilborne Plant Pathogens* (pp. 20-49). St. Paul, MN: The American Phytopathological Society Press.

Campbell, R. and R.M. McDonald. 1989. *Microbial Inoculation of Crop Plants.* Society for General Microbiology. Oxford University Press.

Carson, R. 1962. *Silent Spring.* Boston, MA: Houghton Mifflin.

Chet, I., A. Ordenlich, R. Shapiro, and A. Oppenheim. 1990. Mechanisms of biocontrol of soil-borne plant pathogens by rhizobacteria. *Plant and Soil* 29: 85-92.

Childs, D., T. Jordan, M. Ross, and T. Bauman. n.d. Weed control in no-tillage systems. CT-2, Cooperative Extension Service, Purdue University, <http://www.agcom.purdue.edu/AgCom/Pubs/CT/CT-2.html>.

Clark, A. (Coordinator). 1998. *Managing Cover Crops Profitably,* Second Edition. Sustainable Agriculture Network Handbook Series Number 3. Beltsville, MD: Sustainable Agriculture Network, National Agricultural Library.

Cooke, G.W. 1967. *The Control of Soil Fertility.* New York: Hafner Publishing Co.

Cooperband, L.R. 2000. Sustainable use of by-products in land management. In J.M. Bartels and W.A. Dick (Eds.), *Land Application of Agricultural, Industrial, and Municipal By-Products.* Soil Science Society of America Book Series #6. Madison, WI: Soil Science Society of America.

Costello, M. 1999. Native perennial grasses as cover crops. *Fruit Grower,* September:32.

Creamer, N.G. and K.R. Baldwin. 2000. An evaluation of summer cover crops for use in vegetable production in North Carolina. *HortScience* 35:600-603.

Dahnke, W.C. and G.V. Johnson. 1990. Testing soils for available nitrogen. In R.L. Westerman (Ed.), *Soil Testing and Plant Analysis,* Third Edition. Madison, WI: Soil Science Society of America.

Dalal, R.C. and R.J. Mayer. 1986. Long-term trends in fertility of soils under continuous cultivation and cereal cropping in southern Queensland: III. Distribution and kinetics of soil organic carbon in particle-size fractions. *Australian Journal of Soil Research* 24:293-300.

DePolo, J. 1990. Managing manure in Michigan. *Futures,* Fall, 8:3-11.

Dick, W.A., E.L. McCoy, W.M. Edwards, and R. Lal. 1991. Continuous application of no-tillage to Ohio soils. *Agronomy Journal* 83:65-73.

Donald, W.W. 2000. Timing and frequency of between-row mowing and band applied herbicide for annual weed control in soybean. *Agronomy Journal* 92:1013-1019.

Edwards, J.H. and A.V. Someshwar. 2000. Chemical, physical, and biological characteristics of agricultural and forest by-products for land application. In J.M. Bartels and W.A. Dick (Eds.), *Land Application of Agricultural, Industrial, and Municipal By-Products.* Soil Science Society of America Book Series #6. Madison, WI: Soil Science Society of America.

Edwards, J.H., C.W. Wood, D.L. Thurlow, and E. Ruf. 1992. Tillage and crop rotation effects on fertility status of a hapludult. *Soil Science Society of America Journal* 56:1577-1582.

Eggert, F.P. and C. Kahrmann. 1984. Response of three vegetable crops to organic and inorganic nutrient sources. In D.M. Kral (Ed.), *Organic Farming.* ASA Special Bulletin # 46. Madison, WI: American Society of Agronomy, Crop Science Society of America, Soil Science Society of America.

Elliott, L.F., R.I. Papendick, and J.F. Parr. 1984. Summary of the organic farming symposium. In D.M. Kral (Ed.), *Organic Farming*. ASA Special Publication Number 46. Madison, WI: American Society of Agronomy, Crop Science Society of America, Soil Science Society of America.

Erickson, A.E. and D.M. Van Doren. 1960. The relation of plant growth and yield to soil oxygen availability. *Transactions 7th International Congress Soil Science*, III:428-434.

Fisher, J. 1990. Alternative agriculture: IPM cuts costs of pesticides. *Florida Grower and Rancher* February:28, 29.

Fitzpatrick, G.E., E.R. Duke, and K.A. Klock-Moore. 1998. Use of compost products for ornamental crop production: Research and growers' experiences. *HortScience* 33:941-944.

Flint, M.L. and S.H. Dreistadt. 1998. *Natural Enemies Handbook*. University of California Division of Agriculture and Natural Resources Publication 3386. Berkeley: UC Division of Agriculture and Natural Resources and University of California Press.

Follett, R.H., L.S. Murphy, and R.L. Donahue. 1981. *Fertilizers and Soil Amendments*. Englewood Ciffs, NJ: Prentice-Hall, Inc.

Frye, W.W., J.J. Varco, R.I. Blevins, M.S. Smith, and S.J. Corak. 1988. Role of annual legume cover crops in efficient use of water and nitrogen. In W.L. Hargrove (Ed.), *Cropping Strategies for Efficient Use of Water and Nitrogen*. ASA Special Publication 51. Madison, WI: American Society of Agronomy, Crop Science Society of America, Soil Science Society of America.

Futch, S.H. 2001. Middles management in Florida citrus. *Citrus Industry* October: 19-22.

Garrett, S.D. 1970. *Pathogenic Root-Infecting Fungi*. London: Cambridge University Press.

Gasho, G.T., R.E. Hubbard, T.B. Brenneman, A.W. Johnson, D.R. Summer, and G.H. Harris. 2001. Effects of broiler litter in an irrigated, double-cropped, conservation-tilled rotation. *Agronomy Journal* 93:1315-1320.

Glass, A.D.M. 1989. *Plant Nutrition*. Boston, MA: Jones and Bartlett Publishers, Inc.

Glenn, D.M., W.V. Welker, and G.M. Greene. 1996a. Sod competition in peach production: I. Managing sod proximity. *Journal of the American Society of Horticultural Science* 121:666-669.

Glenn, D.M., W.V. Welker, and G.M. Greene. 1996b. Sod competition in peach production: II. Establishment beneath mature trees. *HortScience* 121:670-675.

Glinski, J. and J. Lipiec. 1990. *Soil Physical Conditions and Plant Roots*. Boca Raton, FL: CRC Press.

Goldstein, N. 2000. The state of biosolids in America. *Biocycle* 41:50-54.

Gooch, J.J. 1997. Frosty fields yield early vegetables. *American Vegetable Grower* 45:39-40.

Goutin, R. 1998. Commercial composting systems. *HortScience* 33:932-933.

Greenland, R.G. 2000. Optimum height at which to kill barley used as a living mulch in onions. *HortScience* 35:853-855.

Grossl, P.R. and W.P. Inskeep. 1991. Precipitation of dicalcium phosphate dihydrate in the presence of organic acids. *Soil Science Society of America Journal* 55:670-675.

Grossman, J. 2001. Fighting insects with living mulches. Ecological Agriculture Projects. McGill University(Macdonald Campus), Ste-Anne-de-Bellevue, Quebec, Canada, <http://www.eap.mcgill.ca/PCMPC_1.htm>.

Grubinger, V. 1997. Ten steps toward organic weed control. *American Vegetable Grower* February:22-24.

Handayanto, E., G. Cadish, and K.E. Giller. 1997. Regulating N mineralization from plant residues by manipulation of quality. In G. Kadish and K.E. Giller (Eds.), *Driven by Nature: Plant Litter Decomposition* (pp. 176-186). Wallingford, UK: CAB International.

Hargrove, W.L. 1990. Role of conservation tillage in sustainable agriculture. In J.P. Mueller and M.G. Wagner (Eds.), *Conservation Tillage for Agriculture in the 1990's* (pp. 28-34). NCSU Bulletin 90-1. Raleigh, NC: Department of Crop Science and Soil Science, North Carolina State University.

Hassink, J., F.J. Mattus, C. Chenu, and J.W. Dadenberg. 1997. Interactions between soil biota, soil organic matter, and soil structure. In L. Brussaard and R. Ferrera-Cerrato (Eds.), *Soil Ecology in Sustainable Agricultural Systems* (pp. 15-35). Boca Raton, FL: CRC Press.

Hauser, S., B. Vanlauwe, D.O. Asawalam, and L. Norgrave. 1997. Role of earthworms in traditional and improved low-input agricultural systems in West Africa. In L. Brussaard and R. Ferrera-Cerrato (Eds.), *Soil Ecology in Sustainable Agricultural Systems* (pp. 113-136). Boca Raton, FL: CRC Press.

Heal, O.W., J.M. Anderson, and M.J. Swift. 1997. Plant litter quality and decomposition: An historical overview. In G. Cadisch and K.E. Giller (Eds.), *Driven by Nature: Plant Litter Quality and Decomposition* (pp. 3-30). Wallingford, UK: CAB International.

Heinz, K.M. 2001. Compatibility of pesticides with biological control agents. *Greenhouse Business*, 7:43-44.

Hillel, D. 1987. *The Efficient Use of Water in Irrigation*. World Bank Technical Paper Number 64. Washington, DC: The World Bank.

Hochmuth, G.J. and E.A. Hanlon. 1995. SP 177, *Commercial Vegetable Crop Nutrient Requirements in Florida*. Gainesville, FL: University of Florida, Institute of Food and Agricultural Sciences.

Hoitink, H.A.J., L.V. Madden, and M.J. Boehm. 1996. Relationships among organic matter decomposition level, microbial species diversity, and soilborne disease severity. In R. Hall (Ed.), *Principles and Practice of Managing Soilborne Plant Pathogens*. St. Paul, MN: The American Phytopathological Society.

Hoitink, H.A.J., A.J. Stone, and D.Y. Han. 1997. Suppression of plant diseases by composts. *HortScience*. 32:184-187.

Hue, N.V., N. Hirunburana, and R.L. Fox. 1988. Boron status of Hawaiian soils as measured by B sorption and plant uptake. *Communications in Soil Science Plant Analyses* 19:517-528.

Izaurralde, R.C., W.B. McGill, J.A. Robertson, N.G. Juma, and J.J. Thurston. 2001. Carbon balance of the Breton classical plots over half a century. *Soil Science Society of America Journal* 65:431-441.

Jenny, H. 1941. *Factors of Soil Formation*. New York: McGraw-Hill.

Jensen, H.L. 1965. Nonsymbiotic nitrogen fixation. In W.V. Batholemew and F.C. Clark (Eds.), *Soil Nitrogen* (pp. 436-480). Madison, WI: American Society of Agronomy.

Julien, M.H. 1992. *Biological Control of Weeds: A World Catalogue of Agents and Their Target Weeds,* Third Edition. Brisbane, Australia: CAB International.

Kasper, T.C., D.C. Erbach, and R.M. Cruise. 1990. Corn response to seed-row residue removal. *Soil Science Society of America Journal* 54:1112-1117.

Lal, R. (Ed.). 1998. *Soil Quality and Agricultural Sustainability*. Chelsea, MI: Ann Arbor Press.

Legere, A., N. Samson, R. Rioux, D.A. Angers, and R.R. Simaud. 1997. Response of spring barley to crop rotation, conservation tillage, and weed management intensity. *Agronomy Journal* 89:628-638.

Leighty, C.E. 1938. Crop rotation. In *Soils and Men. Yearbook of Agriculture* (pp. 406-430). Washington, DC: United States Department of Agriculture.

Letey, J. 1994. Is irrigated agriculture sustainable? In D.M. Kral (Ed.), *Soil and Water Science: Key to Understanding Our Global Environment*. SSSA Special Publication Number 41. Madison, WI: Soil Science Society of America.

Linderman, R.G. 1989. Organic amendments and soil-borne disease. *Canadian Journal of Plant Pathology* 11:180-183.

Loschinkohl, C. and M.J. Boehm. 2001. Composted biosolids incorporation improves turfgrass establishment on disturbed urban soil and reduces leaf rust severity. *HortScience* 36:790-794.

Lynch, J.M. (Ed.). 1990. *The Rhizosphere*. New York: John Wiley and Sons.

Machovej, R.M. and T. Obreza. 1999. Biosolids: Are these residuals the same? *Citrus and Vegetable Magazine* September:14, 16.

Mackowiak, C.L., P.R. Grossl, and B.G. Bugbee. 2001. Beneficial effects of humic acid on micronutrient availability to wheat. *Soil Science Society of America Journal* 65:1744-1750.

Maronek, D.M., J.W. Hendrix, and J. Kiernan. 1981. Mycorrhizal fungi symbiosis and their importance in horticultural crop production. In W.J. Kender, F.J. Morousky, and L.E. Pearce (Eds.), *Horticultural Reviews* 3:172-213.

Meisinger, J.J., V.A. Bandel, G. Stanford, and J.O. Legg. 1985. Nitrogen utilization of corn under minimal tillage and moldboard plow tillage. 1. Four-year results using labeled N fertilizer. *Agronomy Journal* 77:602-611.

Mengel, D.B., D.W. Nelson, and D.M. Huber. 1982. Placement of nitrogen fertilizer for no-till and conventional till corn. *Agronomy Journal* 74:515-518.

Merwin, I., M. Billonen, and J.A. Ray. 1995. Compost mulch, canola cover crops, and herbicides affect soil fertility, apple tree yield, and nutrition. 95th Annual Meeting American Society for Horticultural Science, July 30-August 3. Montreal, Quebec, Canada.

Miller, E.C. 1938. *Plant Physiology.* New York: McGraw-Hill.

Miller, R.W. and R.L. Donahue. 1995. *Soils in Our Environment,* Seventh Edition. Englewood Cliffs, NJ: Prentice-Hall.

Mitchum, W.E. and M.L. Parker. 2000. Managing pecan orchard floors. *Fruit Grower,* January.

The Morrow Plots. 1960. Illinois Agricultural Experiment Station Circular 3777. Urbana, IL: University of Illinois.

Morse, R., T. Elkner, and S. Groff. 2001. *No-Till Pumpkin Production: Principles and Practices.* Pennsylvania Vegetable Marketing and Research Program.

Morse, R.D., D.H. Vaughan, and L.W. Belcher. 1993. Evolution of conservation tillage systems for transplanted crops—Potential role of the subsurface tiller transplanter (SSS-T). In P.K. Bollich (Ed.), *Evolution of Conservation Tillage Systems* (pp. 145-151). Proceedings Southern Conservation Tillage Conference for Sustainable Agriculture, Monroe, LA, June 15-17.

Nafziger, E.D. 1994. Cover crops and cropping systems. *Illinois Agronomy Handbook 1995-1996.* Cooperative Extension Service Circular 1333. University of Illinois at Urbana-Champaign.

Needelman, B.A., M.A. Wander, G.A. Bolero, C.W. Boast, G.K. Sims, and D.G. Bullock. 1999. Interaction of tillage and soil texture: Biologically active soil organic matter in Illinois. *Soil Science Society of America Journal* 63:1326-1334.

Pang, X.P. and J.L. Letey. 2000. Organic farming: Challenge of timing nitrogen availability to crop nitrogen requirements. *Soil Science Society of America Journal* 64:247-253.

Patrick, Z.A. and L.W. Koch. 1958. Inhibition of respiration, germination, and growth by substances arising during the decomposition of certain plant residues in the soil. *Canadian Journal of Botany* 36:621-647.

Paulitz, T.C. 1997. Biological control of root pathogens in soilless and hydroponic systems. *HortScience* 32:193-195.

Peet, M. 2001. Sustainable practices for vegetable production in the south. <http://www.cals.ncsu.edu/sustainable/peet/cover/l-mulch.html>.

Perez-Moreno, J. and R. Ferraro-Cerrado. Mycorrhizal interactions with plants and soil organisms in sustainable agrosystems. In L. Brussaard and R. Ferrera-Cerrato (Eds.), *Soil Ecology in Sustainable Agricultural Systems* (pp. 26-27). Boca Raton, FL: CRC Press.

Phatak, S.H. 1998. Georgia cotton, peanut farmers use cover crops to control pests. In A. Clark (Coordinator), *Managing Cover Crops Profitably.* Sustainable Agricultural Network Handbook Series Book 3. Beltsville, MD: Sustainable Agricultural Network, National Agricultural Library.

Piccolo, A. and J.S.C. Mbagwu. 1999. Rate of hydrophobic components soil organic matter in soil aggregate stability. *Soil Science Society of America Journal* 63:1801-1810.

Piccolo, A., G. Pietramellara, and J.S.C. Mbagwu. 1997. Use of humic substances as soil conditioners to increase aggregate stability. *Geoderma* 75:265-277.

Pierce, F.J. and W. E. Larson. 1998. Developing criteria to evaluate sustainable land management. In J.M. Kimble (Ed.), *Proceedings of 8th International Soil Management Workshop: Utilization of Soil Survey Information for Sustainable Land Use* (pp. 7-14). Lincoln, NE: USDA-SCS National Soil Survey Center.

Pierce, F.L., M.C. Fortin, and M.J. Staton. 1992. Immediate and residual effects of zone tillage in rotation with no-tillage on soil physical properties and corn performance. *Soil Tillage Research* 24:149-165.

Prepare planters for con-till. 2000. "Con-Till Cotton," A Special Supplement to Farm Chemicals, August. Willoughby, OH: Meister Publishing Co.

Proceedings of the 1990 Southern Region Conservation Tillage Conference. July 16-17, 1990, Raleigh, NC. NCSU Special Bulletin 90-1. Raleigh NC: Department of Crop Science and Soil Science, North Carolina State University.

Putnam, A.R. 1990. Vegetable weed control with minimal herbicide inputs. Special insert. *HortScience* 25:155-159.

Raju, P.S., R.B. Clark, J.R. Ellis, and J.W. Maranville. 1987. Vesicular-arbuscular mycorrhizal infection effects on sorghum growth, phosphorus efficiency, and mineral uptake. *Journal of Plant Nutrition* 10:1331-1339.

Randall, G.W. and V.A. Bandel. 1991. Overview of nitrogen management for conservation tillage. In T.J. Logan (Ed.), *Effects of Conservation Tillage on Groundwater Quality, Nitrogen and Pesticides.* Chelsea, MI: Lewis Publishers.

Randall, G.W., J.A. Vetsch, and T.S. Murrell. 2001. Corn response to phosphorus placement under various tillage practices. *Better Crops* 85:12-13.

Rasmussen, P.E., R.W. Rickman, and B. Klepper. 1997. Residue and fertility effects on yield of no-till wheat. *Agronomy Journal* 89:563-567.

Reganold, J.P., L.F. Elliott, and Y.L. Unger. 1987. Long-term effects of organic and conventional farming on soil erosion. *Nature* 330:370-372.

Reiners, S. and O. Wickerhauser. 1995. The use of rye as a living mulch to control weeds in bell pepper production. 92nd Annual Meeting American Society for Horticultural Science, July 30-August 3. Montreal, Quebec, Canada.

Reynolds, M.P., K.D. Sayre, and H.E. Vivar. 1994. Intercropping wheat and barley with N-fixing legume species: A method for improving ground cover, N-use efficiency and productivity in low-input systems. *Journal of Agricultural Science* 23:175-183.

Robbins, G.J. (Ed.). 1989. *Alternative Agriculture.* Washington, DC: National Research Council, National Academy Press.

Roe, N. 1998. Compost utilization for vegetable and fruit crops. Special insert. *HortScience* 33:934-937.

Roe, N.E., P.J. Stoffella, and H.H. Bryan. 1994. Growth and yields of bell pepper and winter squash with organic and living mulches. *Journal American Society of Horticultural Science* 122:433-437.

Roe, N.E., P.J. Stofella, and D. Graetz. 1997. Composts from various municipal solid waste feedstocks affect vegetable crops. 1. Emergence and seedling growth. *Journal of the American Society of Horticultural Science* 122:427-432.

Roe, N.E., H.H. Bryan, P.J. Stoeffela, and T. Winsberg. 1992. Use of compost as mulch on bell peppers. *Proceedings Florida State Horticultural Society* 105:336-338.

Roka, F.M., R.M. Muchovej, and T.A. Obreza. 1999. Assessing the economic value of biosolids. *Citrus and Vegetable Magazine* October:10-14.

Ross, S.M., J.R. King, R.C. Izaurralde, and J.T. O'Donovan. 2001. Weed suppression by seven clover species. *Agronomy Journal* 93:820-827.

Russell, E.W. 1961. *Soil Conditions and Plant Growth.* London: Longmans, Green.

Salter, R.M. 1933. Factors affecting the accumulation and loss of nitrogen and organic in cropped soils. *Journal of the American Society of Agronomy* 25:622-630.

Schlegel, A.J. and J.L. Havlin. 1997. Green fallow for the central Great Plains. *Agronomy Journal* 89:762-767.

Senn, T.L. and A.R. Kingman. 1973. *A Review of Humus and Humic Acids.* Research Series # 145. Clemson, SC: South Carolina Agricultural Experiment Station.

Servis, R. 1992. Ag plastic recycling on horizon. *Florida Grower and Rancher* 85:40.

Seyfried, M.S. and M.D. Murdock. 2001. Response of new soil water sensors to variable soil water content, and temperature. *Soil Science Society America Journal* 65:28-34.

Shangning, J. and P.W. Unger. 2001. Soil water accumulation under different precipitation, potential evaporation, and straw mulch conditions. *Soil Science Society of America Journal* 65:442-448.

Sikoura, L.J. and D.E. Stout. 1996. Soil organic carbon and nitrogen. In *Methods for Assessing Soil Quality* (pp. 157-167). Soil Science Society of America Special Publication Number 49.

Sims, J.T. and J.T. Wolf. 1994. Poultry manure management. Agricultural and environmental issues. *Advances in Agronomy* 52:1-83.

Six, J., E.T. Elliott, and K. Paustian. 1999. Aggregate and soil organic matter dynamics under conventional and no-tillage systems. *Soil Science Society of America Journal* 63:1350-1358.

Sprague, M.A. and G.B. Triplett (Eds.). 1986. *No-Tillage and Surface Tillage Agriculture.* New York: John Wiley and Sons.

Stephens, C.T. 1990. Minimizing pesticide use in a vegetable management system. *HortScience* 25:164-168.

Stoffella, P.J., Y. Li, N.E. Roe, M. Ozores-Hampton, and D.A. Graetz. 1997. Utilization of composted wastes in vegetable production systems. Food and Fertilizer Technology Center, Taipei City, Taiwan. Technical Bulletin 147.

Sullivan, D.M., A.I. Bary, D.R. Thomas, S.S. Fransen, and C.G. Cogger. 2002. Food waste compost effects on fertilizer nitrogen efficiency, available nitrogen, and tall fescue yield. *Soil Science Society of America Journal* 66:154-161.

Sutton, A.J., D.D. Jones, B.C. Joern, and D.M. Huber. 1994. *Animal Manure As a Plant Nutrient Resource*. Purdue University Cooperative Extension Service Bulletin ID–101. West Lafayette, IN: Purdue University.

Teasdale, J.H. and A.A. Abdul-Baki. 1995. Soil temperature and tomato root growth under black polyethylene and hairy vetch mulches. 95th Annual Meeting, American Society for Horticultural Science, July 30-August 3, Montreal, Quebec, Canada.

Thapa, B.B., D.B. Garrity, D.K. Cassel, and A.R. Mercado. 2000. Contour grass strips and tillage affect corn production on Philippine steepland oxisols. *Agronomy Journal* 92:98-105.

Thomashow, L.S. and D.M. Weller. 1996. Molecular basis of pathogen suppression by antibiosis in the rhizosphere. In R. Hall (Ed.), *Principles and Practice of Managing Soilborne Plant Pathogens* (pp. 80-103). St. Paul, MN: American Phytopathological Society.

Thompson, L.M. and F.R. Troeh. 1978. *Soils and Soil Fertility*, Fourth Edition. New York: McGraw-Hill.

Throckmorton, R.L. 1986. Tillage and planting equipment for reduced tillage. In M.A. Sprague and G.B. Triplett (Eds.), *Tillage and Surface-Tillage Agriculture* (pp. 59-91). New York: John Wiley and Sons.

Tian, G., L. Brussard, B.T. Kang, and M.J. Swift. 1997. In G. Kadish and K.E. Gilbert (Eds.), *Driven by Nature: Plant Litter Decomposition* (pp. 125-134). Wallingford, UK: CAB International.

Tinker, P.B. 1980. The role of rhizosphere organisms in phosphorus uptake by plants. In F.E. Khasawneh, E.C. Sample, and E.J. Kamprath (Eds.), *The Role of Phosphorus in Agriculture* (pp. 617-654). Madison, WI: American Society of Agronomy, Crop Science Society of America, Soil Science Society of America.

Tisdale, S.L. and W.L. Nelson. 1975. *Soil Fertility and Fertilizers*. Third edition. New York: Macmillan.

Triplett, G.B. and A.D. Worsham. 1986. Principles of weed management with surface-tillage systems. In M.A. Sprague and G.B. Triplett (Eds.), *No-Tillage and Surface Tillage Agriculture* (pp. 319-347). New York: John Wiley and Sons.

Tropical kudzu, maize and peach palms revive Peruvian soils. 1988. *International Ag-Sieve*, 1(3). Emmaus, PA: Rodale International.

Varsa, E.C., D. Rovey, and G. Kapusta. 1995. Effect of residue density, strip tillage, and starter K on no-till corn grown in wheat stubble. In G. Rehm (Ed.), *Proceed-*

ings North Central Extension-Industry Soil Fertility Conference. St. Louis, MO (pp. 81-87). Manhattan, KS: Potash and Phosphate Institute.

Vyn, T.J., J.G. Faber, K.J. Janovicek, and E.G. Beauchamp. 2000. Cover crop effects on nitrogen availability to corn following wheat. *Agronomy Journal* 92: 915-924.

Waksman, S.A. 1938. *Humus,* Second Edition, Revised. Baltimore, MD: Williams and Wilkins.

Wallace, A. and G.A. Wallace. 1994. A possible flaw in EPA's 1993 new sludge rule due to heavy metal interactions. *Communications Soil Science Plant Analysis* 25:129-135.

Webley, D.M. and R.B. Duff. 1965. The incidence in soils and other habitats of microorganisms producing alpha-ketogluconisc acid. *Plant and Soil* 22:307-313.

Whitworth, J. 1995. Weed management methods influence growth of 'Navaho' blackberries. 95th Annual Meeting, American Society for Horticultural Science, July 30-August 3. Montreal, Quebec, Canada.

Wolf, B. 1982. An improved universal extracting solution and its use for diagnosing soil fertility. *Communications Soil Science Plant Analyses* 13:1005-1033.

Wolf, B. 1996. *Diagnostic Techniques for Improving Crop Production.* Binghamton, NY: The Haworth Press.

Wolf, B. 1999. *The Fertile Triangle: The Interrelationship of Air, Water, and Nutrients in Maximizing Soil Productivity.* Binghamton, NY: The Haworth Press.

Wolf, B., J. Fleming, and J. Batchelor. 1980. *Liquid Fertilizer Manual.* Peoria, IL: National Fertilizer Solutions Association.

Wolf, B., J. Fleming, and J. Batchelor. 1985. *Fluid Fertilizer Manual.* Manchester, ND: National Fertilizer Solutions Association.

Worsham, A.D. 1990.Weed management strategies for conservation tillage in the 1990's. In J. P. Mueller and M.G. Wagner (Eds.), *Conservation Tillage for Agriculture in the 1990's.* Special Bulletin 90-1. Raleigh, NC: N. C. State University.

Wright, D.L., F.M. Rhoads, and R.L. Stanley Jr. 1980. High level of management needed on irrigated corn. *Solutions* May/June:24-36.

Yermiyahu, U., R. Koren, and Y. Chen. 2001. Effect of composted organic matter on boron uptake by plants. *Soil Science Society of America Journal* 65:1436-1441.

Zandstra, B.H. and D.D. Warncke. 1993. Interplanted barley and rye in carrots and onions. *HortTechnology* 2:214-218.

Index

Page numbers followed by the letter "f" indicate figures; those followed by the letter "t" indicate tables.

Allelopathy, 40-41, 222, 246, 277
  beneficial effects of, 40, 157, 158, 246, 276, 277, 279
  reducing harmful effects of, 40, 41, 146
Alternative agriculture
  basis for a sustainable agriculture, 8
  evaluation of, 8-10
  management practices, 10, 11
Antibiotics, 127
  role of actinomycetes, 116

Biosolids. *See* Sewage, including effluent and biosolids
Bird damage, 283
Buffering, 59
Bulk density of soil, 91-93, 92t

Carbon dioxide, 28-29
Carbon/nitrogen (C:N) ratio
  nitrogen deficiencies and, 39-40, 56-57
  nitrogen fertilization, 65-66
  of organic materials, 21t-22t
  organic matter decomposition, 20-22, 55-57
Cation exchange capacity (CEC), 25, 58
  percentage saturation of cations, 307
Chelates, chelation, 49, 51-52
Compaction of soil, 93-95, 236-237, 246
  and nematode activity, 111
Compost
  benefits of, 130, 185-187, 294

Compost *(continued)*
  composition, 179t
  disease suppression by, 130, 294
  loamless composts, 183-184
  as mulch, 223
  mulches for disease suppression, 302
  preparation, 180-183
  sources, 178-179
Conservation tillage
  benefits of, 131, 228-234
  changes in pH, 257-259
  deleterious effects of, 235-237, 241-247
  equipment for, 249-257, 251f-256f, 307-311, 309f-310f
  fertility, 256-263
  forms of, 227-228, 237-241
  no-tillage, 234-237
    effect on compaction, 236, 237
    effect on soil moisture, 236
    effect on stratification, 237
  plant pests and, 242-245, 263-265, 272-274, 283-284
  reduced costs, 234
  reduced tillage, 237-241, 315
  ridge, 239-241
  zone or strip, 238
Conservation tillage + organic matter additions + mulch (COM)
  adding organic matter for, 288-297, 314
  adding water for, 305-307, 317
  basic approach, 285-312, 316
  controlling pests in, 298-304, 316
  equipment for, 307-312
  supplying food and fiber, 287
  supplying needed nutrients for, 297-298

Cover crops
    for increasing soil organic matter, 137-139, 289-290, 296
    for killed mulches, 215
    mixtures, 155-161
    for nutrient availability, 146-150
    nutrient retention and supply, 148-150
    for pest control, 131, 150-152
    reseeding, 154
    in rotations, 144-145, 152-154
    suppression of, 212
    terminating, 249, 296, 307, 308
    types of, 154-162
    for weed control, 134, 134t, 161, 276-280
Crop residues. *See* Plant residues
Crop rotation
    for disease control, 130, 159, 269-271, 270t-271t
    for increasing soil organic matter, 139-140, 290, 316
    for pest control, 141-143
    using cover crops, 152-153
    for weed control, 275

Disease. *See also* Pests of plants
    as affected by compost mulch, 302
    as affected by conservation tillage, 243
    as affected by moisture and temperature, 117
    as affected by plant residues, 267-269
    computer forecasting systems and, 302
    control of, 269-272, 301-303
    favored by mulch, 224
    resistance, 131
    suppression of, 132

Earthworms, 106, 109-110, 128
Equipment. *See also* Machinery, and intensive agriculture
    chisel plows, 254, 256, 310
    for conservation tillage, 249-256, 316

Equipment *(continued)*
    coulters, 234, 252-253
    effect of tillage tools on residue spread, 218t, 219t, 220t
    planters, 251-253
    sprayers, 255
    stalk choppers, 254, 311
    transplanter, 255, 257
Erosion, 4, 102-104, 232-235, 313

Fertility. *See* Soil fertility
Fertilizers
    activated sewage sludge, 188-189
    artificial
        criticism of, 5, 6, 313
        as a means of increasing yields, 315
    high analysis, 6
    micronutrients in, 6-7
    organic, 5-6
    pollution, 7-8
    reduction in use, 10-11
    synthetic, 8
Food by-products, 192-196
Fungicides
    effect on beneficials, 272
    use of, as affected by conservation tillage, 272

Health of soil, 118-119
Heavy metals in sewage effluents and biosolids, 159-191
Herbicides, 208-210, 212, 249-250, 282-283, 304
Humus, 23-26

Infiltration and percolation, 95-98, 230
Insecticides
    and beneficials, 265-267
    reduced by use of natural organic mulch, 225
    reduced with conservation tillage, 246

Insects
  beneficial
    effect of pesticides on, 265, 267, 272
    for controlling insect pests, 264, 265
    increased with conservation tillage, 242, 265, 300
    as influenced by cover crops, 265, 301
    as influenced by mulch, 224-226
  harmful
    as affected by reduced tillage, 242-243
    controlled by GM crops, 267
Integrated pest management (IPM)
  cost of, 300
  need for, 298-301, 304
Intensive agriculture
  components of, 1-2
  criticisms of, 313
  efficiency of, 2
  food production by, 17
  improving crop yields, 315
  problems, 2-3, 9-11, 313, 315
    correcting problems, 3-4
  soil organic matter and, 3

Legumes, 20, 60, 61t, 67, 68, 68t, 69t, 120
  cover crop, 90, 146, 149, 150, 155, 158-160
  living, 207

Machinery, and intensive agriculture, 1, 2. *See also* Equipment
Macroorganisms of the soil. *See also* Earthworms
  anthropods, 107
  insects, 108-109
  mollusks, 109
  vertibrates, 107
Manure
  nutrient release, 66-67, 176-177
  nutrient content, 171-176, 172t-173t
  production of, 170-171, 171t, 293-294
  utilization of, 291-294

Micorrhizal fungi, 61-63, 115-116, 122-123
Microorganisms of soil. *See also* Micorrhizal fungi; Nematodes
  actinomycetes, 116
  algae, 111
  bacteria, 116-118, 123-125
  fungi, 112-116, 123-125, 128
  protozoa, 112
Moisture holding capacity (MHC), 99-102, 100t, 305
Monoculture, 142, 313
Mulch
  associated pests, 223-224
  comparative value of natural organic and plastic, 205
  crop residues, 215
  deleterious effects of, 217-223
  disposing of dry organic materials, 38-39
  killed mulches, 213-214
  living mulches, 205-213, 297
  maximizing effects of, 215-217, 218t-220t
  organic mulches, 201-202, 204
  pest protection by, 224-226
  plastic, 200-201, 205
  soil moisture, 202-203, 203t
  soil structure, 204-205
  soil temperature, 203, 221, 222
  stubble mulch, 238
  weed control, 275-276
Mycorrhizae
  benefits of, 62, 63
  effect on available nutrients, 122, 123
  types of, 61, 62

Nematodes, 110-111
  role of conservation tillage on, 272-273
Nitrogen
  conservation tillage and, 259-261
  fixation, 60, 61t, 111, 119-122, 146-149
  release from organic matter, 69-78
Nonlegumes, 146, 148, 149, 155-158, 161
No-tillage. *See* Conservation tillage

Nutrients, plant
  essential, 45-46
  and plant residues, 47t, 49t-54t
  release from plant residues, 54-57
  release from soil organic matter, 69-78, 69t-70t, 72t-73t

Organic farming, 8-11, 286-287, 315
  as basis for sustainable agriculture, 314
  failure to supply sufficient food and fiber, 314
Organic matter (OM). *See also* Soil organic matter (SOM)
  adding
    as animal and plant wastes, 192-197, 295, 296
    as composts, 178-187, 294
    by growing it in place, 137-162, 288-291
    incorporating versus surface placement, 199-200, 202-205
    as manure, 170-178, 291-294
    as peats, 191-192, 295
    as plant residues, 162-163, 163t, 288-290
    as sewage effluents and biosolids, 294
    as wood by products, 296
  allelopathy, 40-41
  biological effects of, 105-135
  chelation of micronutrients, 49, 51, 52
  C/N ratio, 20-22, 39-40
  decomposition, 19-22, 23t, 26-33
  excess, 37-39
  effect of, on soil organisms, 105-118
  fragile versus nonfragile, 22, 23
  heavy metals, 41, 189-190, 190t
  humus or SOM from, 23-26
  nutrient availability, effect of organic matter on, 57-61
  nutrients in, 45-57, 64-67, 297-298
    as a basis for changing fertilizer recommendations, 64-78
    imbalances of, 42
    losses of, 63, 64
    release of, 54-57, 64-76
    variability of, 48, 49

Organic matter *(continued)*
  persistence, 22-23
  pests, 42-43
  physical effects of, 79-104
    aggregate stability, 98, 99
    bulk density, 91-93
    moisture-holding capacity (MHC), 99-102
    soil compaction, 93-95
    soil porosity, 84-91
    soil structure, 81-84
    water infiltration and percolation, 93-98
  placement, 199-226
  salts, 42
  soil textural classes and, 27
  sustaining agriculture, 14-16
  sustaining soils, 12-14
Oxygen, soil, 26-28, 37-38, 71

Parasitism, 128-129
Peat, 191-192, 193t, 295
Percolation. *See* Infiltration and percolation
Pest control
  with COM system, 298-304
  cover crops as aids for, 150-152
  diseases, 129-132, 150-151, 269-273
  handling crop residues for, 165-167
  with mulch, 224-226
Pesticides
  application equipment, 311-312, 312f
  beneficial insects and, 265-267, 272, 316
  elimination of, 316
  EPA, 5
  FIFRA, 4, 313
  from mulch, 222-223
  need for, 271-272
  reduction in use, 10
  regulation, 4-5, 313
Pests of plants
  as affected by conservation tillage, 243, 267-269
  as affected by cover crops, 150-152
  effect of tillage on, 241-247

Pests of plants *(continued)*
    insects as, 132-133, 264-267
        as affected by conservation
            tillage, 242-243
    nematodes as, 135, 244-245
        as affected by conservation
            tillage, 244-245, 245t
    suppression of, 125-132
    weeds as, 133-134, 273-276,
        280-283
        as affected by conservation
            tillage, 243-244
Plant residues
    amounts, 162-165
    benefits, 164-165
    handling, 165-167
    planting through, 310
Pollution, 8, 12, 287, 303
Porosity of soil
    and bulk density, 93
    classification, 85-86
    effect of OM, 58, 230
    importance of, 84-89
    maintenance of, 90-91

Quality of soil, 118-119

Reduced tillage. *See* Conservation tillage
Rotations. *See* Crop rotation

Salts
    as affected by organic matter, 317
    introduced by irrigation, 305-307,
        317
    reduction of plant growth, 305
SAR, 306
Sewage, including effluent and
        biosolids
    agricultural use, 190-191
    heavy metals, 189-190, 190t, 295
    as a source of organic matter, 294,
        295
    treatment, 187-189
Slug damage, 284
Small animal damage, 224, 284

Soil actinomycetes, 116
Soil aggregates
    effect of conservation tillage, 230
    effect of organic matter, 118, 124
Soil algae, 111
Soil bacteria, 116-118
Soil fertility
    as affected by conservation tillage,
        256-263, 297, 298
    as affected by organic matter,
        118-125
Soil fungi, 112-116
    parasites, 114, 115
    symbiotic forms, 115-116
Soil insects, 108-109. *See also* Insects
Soil organic matter (SOM)
    agricultural productivity, 14, 33-35
    for controlling weeds, 133-134
    formation of, 106, 108
    humus, 23-26
    increasing
        by adding organic matter, 35-37,
            137-167, 177-178
        by reduced tillage, 167, 227-228
    infiltration and percolation of water
        due to, 95-98
    intensive agriculture and, 3
    losses of, 33-36
    microorganisms, 12-13
    nitrogen, influence of, 32-33
    problems associated with, 37-43,
        294-295
    soil formation and, 12-13
    soil structure, 13-14, 98-99
    sustainable soil, 12, 14-15
    tillage and, 3, 8, 11, 38-39, 227-228
Soil oxygen, 231
Soil protozoa, 112
Soil testing, need for, 7, 8, 78, 316
Stratification, 237, 247
Subsoiling, 308, 309
Summary, 313-317
Sustainable agriculture, 11-12, 15-17,
    166, 287, 289, 307

Tillage. *See also* Conservation tillage
    cost of, 234
    effect on plant residues, 164
    erosion and, 232-234

Tillage *(continued)*
    mulch and, 228
    no-till, 234-237
    ridge tillage, 239-241
    soil organic matter and, 3, 8, 11, 38-39, 227-228
    soil structure affected by, 230, 231t
    soil temperature affected by, 229
    strip tillage, 238-239
    zone tillage, 238-239

Waste products
    animal, 192
    composting of. *See* Compost
    food, 192-196, 295-296
    manure. *See* Manure

Waste products *(continued)*
    sewage. *See* Sewage, including effluent biosolids
    wastewater, 187-191, 195-196
    wood, 196-197, 197t, 296
Water holding capacity. *See* Moisture holding capacity (MHC)
Water use, 305, 317
Weeds
    as affected by reduced tillage, 273-274
    control of by biological agents, 274, 275, 280, 281t-282t
    control of by herbicides, 282-283
    suppression by cover crops, 276, 277
    suppression by mulch, 277

SPECIAL 25%-OFF DISCOUNT!
Order a copy of this book with this form or online at:
http://www.haworthpressinc.com/store/product.asp?sku=4805

SUSTAINABLE SOILS
The Place of Organic Matter in Sustaining Soils and Their Productivity

_____ in hardbound at $52.46 (regularly $69.95) (ISBN: 1-56022-916-0)

_____ in softbound at $37.46 (regularly $49.95) (ISBN: 1-56022-917-9)

Or order online and use Code HEC25 in the shopping cart.

COST OF BOOKS_____

OUTSIDE US/CANADA/
MEXICO: ADD 20%_____

POSTAGE & HANDLING_____
(US: $5.00 for first book & $2.00
for each additional book)
Outside US: $6.00 for first book
& $2.00 for each additional book)

SUBTOTAL_____

IN CANADA: ADD 7% GST_____

STATE TAX_____
(NY, OH & MN residents, please
add appropriate local sales tax)

FINAL TOTAL_____
(If paying in Canadian funds,
convert using the current
exchange rate, UNESCO
coupons welcome)

☐ **BILL ME LATER:** ($5 service charge will be added)
(Bill-me option is good on US/Canada/Mexico orders only;
not good to jobbers, wholesalers, or subscription agencies.)

☐ Check here if billing address is different from
shipping address and attach purchase order and
billing address information.

Signature_____

☐ **PAYMENT ENCLOSED:** $_____

☐ **PLEASE CHARGE TO MY CREDIT CARD.**

☐ Visa ☐ MasterCard ☐ AmEx ☐ Discover
☐ Diner's Club ☐ Eurocard ☐ JCB

Account # _____

Exp. Date_____

Signature_____

Prices in US dollars and subject to change without notice.

NAME_____
INSTITUTION_____
ADDRESS_____
CITY_____
STATE/ZIP_____
COUNTRY_____ COUNTY (NY residents only)_____
TEL_____ FAX_____
E-MAIL_____

May we use your e-mail address for confirmations and other types of information? ☐ Yes ☐ No
We appreciate receiving your e-mail address and fax number. Haworth would like to e-mail or fax special discount offers to you, as a preferred customer. **We will never share, rent, or exchange your e-mail address or fax number.** We regard such actions as an invasion of your privacy.

Order From Your Local Bookstore or Directly From
The Haworth Press, Inc.
10 Alice Street, Binghamton, New York 13904-1580 • USA
TELEPHONE: 1-800-HAWORTH (1-800-429-6784) / Outside US/Canada: (607) 722-5857
FAX: 1-800-895-0582 / Outside US/Canada: (607) 722-6362
E-mail to: getinfo@haworthpressinc.com
PLEASE PHOTOCOPY THIS FORM FOR YOUR PERSONAL USE.
http://www.HaworthPress.com